Fluid Power
Educational
Series

Hydraulic Linear Actuators

(In the English Units)

Joji Parambath

Hydraulic Linear Actuators
(In the English Units)

Copyright © 2026 Joji Parambath

ISBN: 9798653731228

https://jojibooks.com

First Edition: 2020
Revision: 2022
Revision: 2026

Disclaimer of Liability

The contents of this book have been checked for accuracy. Since deviations cannot be precluded entirely, we cannot guarantee full agreement. Only qualified personnel should be permitted to install and service pneumatic and hydraulic equipment. Qualified persons are defined as persons who are authorized to commission, ground, and tag circuits, equipment, and systems following established safety practices and standards.

Table of Contents

PREFACE

Hydraulic cylinders are simple, low-cost, and easy to install, making them ideal for generating powerful linear motion. Manufacturers are introducing a range of actuators with innovative features to improve reliability, efficiency, and safety. The latest industrial hydraulic cylinders can incorporate sensor feedback and electro-hydraulic servo valves for sophisticated speed control and position control of the associated loads.

The book aims to present essential technical information on hydraulic cylinders in a simple, easy-to-understand manner. The topics are logically arranged for a simple-to-complex level progression of the subject matter. Even as technological advancement shows no signs of slowing, a strong foundation in hydraulic technology will help readers quickly understand future developments in the field. The book uses the English system of units.

Many other fluid power topics are covered in other textbooks in the same author's fluid power educational series. A list of all the books is given at the end of the book. Also, please see the details at: https://jojibooks.com.

Enjoy reading the book.
Your feedback is most welcome.

JOJI Parambath

Chapter 1 | Introduction and Basic Cylinder Working

A hydraulic actuator is a positive-displacement device used in a hydraulic system to drive the attached load and perform useful work. Its primary function is to convert hydraulic power into mechanical power. The resulting output motion can be either linear or rotary. Accordingly, there are two basic types of hydraulic actuators. They are: (1) Linear actuators and (2) Rotary actuators. The linear actuators convert hydraulic energy into straight-line mechanical energy. Next, the rotary actuators convert hydraulic energy into rotary mechanical energy. An example of a linear actuator is a cylinder, and the rotary actuator is a hydraulic motor.

Linear Actuators

A linear hydraulic actuator, as shown in Figure 1.1, converts hydraulic power into a controllable linear force or motion or both. Technically and economically, hydraulic and pneumatic cylinders are the optimum form of linear actuators. However, a cylinder in a high-pressure hydraulic system can deliver much higher force than a cylinder of the same size in a pneumatic system.

Figure 1.1 | Hydraulic cylinder

In general, cylinders make up the majority of actuators used in fluid power systems. Hydraulic actuators are available in a range of sizes, types, body styles, and mounting configurations to meet the varied requirements of applications across industrial, aerospace, automotive, and press operations.

Working Principles of a Hydraulic Cylinder

Figure 1.2 shows the cross-sectional view of a typical hydraulic cylinder. It primarily consists of a barrel, piston rod assembly, end caps, and necessary seals and ports. The end caps are securely fastened to the barrel. The piston, with a tight seal, forms two fluid chambers (piston chamber and piston rod chamber) and moves within the barrel, supported by two bearing surfaces. The cylinder has two ports for the system fluid to enter or exit. They are: (1) piston-side (cap-end) port X and (2) rod-side (head-end) port Y.

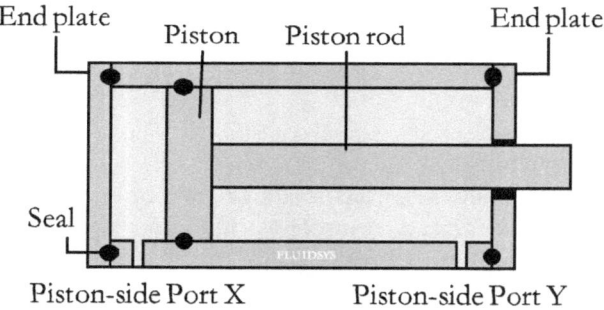

Figure 1.2 | A schematic diagram of a basic hydraulic cylinder

If the system fluid is pumped into the piston chamber through the port X and the fluid in the piston rod chamber is discharged through the port Y, then the piston-and-rod assembly extends with a definite force (thrust). If the system fluid is pumped into the piston rod chamber through the port Y and the fluid in the piston chamber is discharged through the port X, then the piston-and-rod assembly retracts with a definite force (pull). This type of hydraulic cylinder is called a 'double-acting' cylinder because its extension and retraction are achieved hydraulically.

If the motion of the hydraulic cylinder is obtained hydraulically in only one direction, then such a cylinder is called a 'single-acting' cylinder. A spring, gravity, or another external force can then reverse the motion.

2

Chapter 2 | Terms and Definitions
-Hydraulic Cylinders

Some essential parameters for the operation and applications of hydraulic cylinders include bore diameter, piston rod diameter, force (thrust and pull), stroke length, speed, and piston rod buckling. The following sections describe these terms:

Maximum Operating Pressure (P): The pressure that overcomes all resistances in the system, including both useful work and losses, is the maximum operating pressure. Alternatively, it is the maximum working pressure the cylinder can withstand without adverse effects.

Bore Diameter (D): It refers to the diameter at the bore of the cylinder (See Figure 2.1). It can be used to calculate the cylinder's bore area. It is also equal to the piston diameter in a close-fitting hydraulic cylinder.

Piston rod Diameter (d): It refers to the diameter of the piston rod of the cylinder (See Figure 2.1).

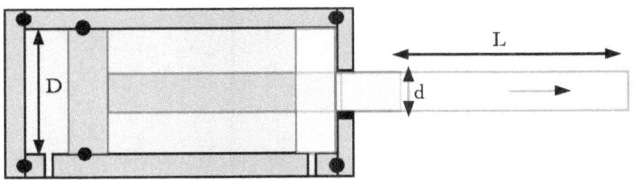

Figure 2.1 | Cylinder parameters

Stroke Length (L): It is the distance through which the piston-and-piston rod assembly of the cylinder moves through the cylinder.

Note: Details of standard bore diameters and piston rod diameters are given in Appendix 1.

Maximum Stroke Length: It is the maximum linear movement that a cylinder can produce. For standard double-acting cylinders, the maximum stroke length is 6.5 ft; for special designs, it can be up to 20 ft.

Cylinder Thrust/Pull (F): The theoretical thrust (F) during the forward stroke or pull (F) during the return stroke of the cylinder can be determined by multiplying the effective area (A) of the piston by the working pressure (P) to which it is subjected, according to Pascal's law.

The active area (A_{ext}), considered for the calculation of the cylinder thrust, is the full area (A_p) of the cylinder bore and is given by ($\pi.D^2/4$).

The parameter 'D' denotes the piston (bore) diameter. Further, the active area (A_{ret}), considered for the calculation of the cylinder pull, is the area (A_p) of the piston minus the piston rod area (A_r), and it is given by the expression [$\pi. (D^2 - d^2)/4$].

The parameter 'd' denotes the piston rod diameter. The theoretical thrust and the theoretical pull are given by:

Thrust, F (lb)	= P (psi) x A_{ext} (in^2)
Pull, F (lb)	= P (psi) x A_{ret} (in^2)

Where,

A_p is the piston area
A_r is the piston rod area
A_{ext} is the active area during extension: ($A_{ext} = A_p$)
A_{ret} is the active area during retraction: ($A_{ret} = A_p - A_r$)

Table A2.1 in Appendix 2 gives the theoretical forces of hydraulic cylinders in the English system of units. These figures do not account for the seal or packing friction in these cylinders. This type of friction is estimated to reduce the cylinders' thrust by about 10%.

Limitations on Maximum Thrust Force

The maximum thrust a cylinder can provide is limited by its piston rod diameter and overall length. In cylinders with longer piston rods, the piston rod must be capable of handling the thrust forces generated by the application. The cylinder must also be supported adequately.

Note that a head-end mounting provides greater column strength than the cap-end mounting, due to the smaller distance between the mounting points in the head-end mounting than that in the cap-end mounting.

The piston rod size of a hydraulic cylinder can be selected from the size charts using its free buckling length and the load imposed on the cylinder.

Example 2.1: A high-pressure double-acting hydraulic press cylinder with an effective piston area of 11 in² for push stroke, and a piston rod area of 3.41 in², operating at 10000 psi, produces what theoretical forces for the push stroke and pull stroke?

Solution

Effective piston area, push stroke, A_{push} = 11 in²
Piston Rod area, A_{rod} = 3.41 in²
Pressure, P   =10000 psi

Effective piston area, pull stroke, $A_{pull} = A_{push} - A_{rod}$
                = $(11 - 3.41)$ = 7.6 in²

Thrust, F_{push} = P x A_{push}
                = (10000) x (11)
                = 110000 lb

Pull, F_{pull} = P x A_{pull}
                = (10000) x (7.6)
                = 76000 lb

Cylinder Input Power: The input power (P_{input}) supplied to the hydraulic cylinder, in the English system units, is given below:

$$P_{input} \text{ (hp)} = P \text{ (psi) x } Q_A \text{ (gpm)}/1714$$

Cylinder Output Power: The mechanical output power (P_{output}) of the hydraulic cylinder in the English system of units is given below:

$$P_{output} \text{ (hp)} = \text{Force (lb) x Velocity (ft/s) }/550$$

The input power supplied must exceed the required output power to account for losses from heat, friction, and leakage.

Cylinder Speed: Assume that the piston rod assembly of a cylinder moves with a velocity of 'v' when pushed by the system fluid with a flow rate 'Q'. Further, assume that the cylinder piston of area 'A' has moved a distance 'S' in time 't' to attain the velocity v.

Figure 2.2 (a) shows the working position of the cylinder with the piston in position 1. Figure 2.2 (b) shows the working position of a cylinder with the piston in position 2, with position 1 superimposed. Mathematically,

$$v = S/t \quad \text{or} \quad t = S/v$$

We can easily relate the theoretical flow rate (Q) of the system fluid to the speed (v) at which the piston rod moves if we consider the cylinder volume (V) that must be filled with the fluid and the distance (S) through which the cylinder piston must travel at the specified speed.

The volume (V) of the cylinder is the length of the stroke (S) multiplied by the piston area (A). The following section provides the flow rate (Q) required to achieve the required speed (v) in English units.

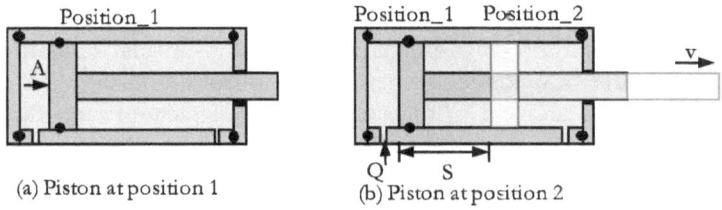

(a) Piston at position 1 (b) Piston at position 2

Figure 2.2 | Illustration of a cylinder in two piston positions

$$Q \text{ (ft}^3/\text{s)} = A \text{ (ft}^2) \times v \text{ (ft/s)}$$

From the equation above, the speed (v) of a given cylinder depends on the flow rate (Q) of the system fluid. That is, a small-diameter cylinder moves faster than a large-diameter cylinder at the same flow rate.

Examination of this equation also shows that the double-acting cylinder tends to run slightly faster during retraction than during extension, provided the flow rate remains constant. This speed difference is mainly due to differences in the active areas exposed to the system fluid.

Cylinder Drift
Under ideal conditions, a hydraulic cylinder should hold its position when stopped.

However, it tends to drift from its intended position. The drift is often attributed to the control valve rather than the cylinder itself. If the valve spool is fully blocked in its neutral position, the cylinder should remain in its current position.

But if the valve spool permits leaks internally in its neutral position, possibly due to a damaged seal, both working ports of the valve may be connected to the system reservoir, and consequently, the cylinder may drift from its intended position. Unintended movement of hydraulic cylinders may cause injuries and accidents.

Operating Temperature

Seal compounds (NBR) are designed for the normal operating temperatures of hydraulic systems, ranging from -5°F to +175°F. Temperatures above 175°F accelerate seal degradation and fluid breakdown; for high operating temperatures up to 390°F, special high-temperature seals (FKM) must be used.

Types of Hydraulic Loads

There are three types of loads associated with hydraulic systems. They are: (1) Resistive load (Positive load), (2) Overrunning load (Negative load), and (3) Inertial load.

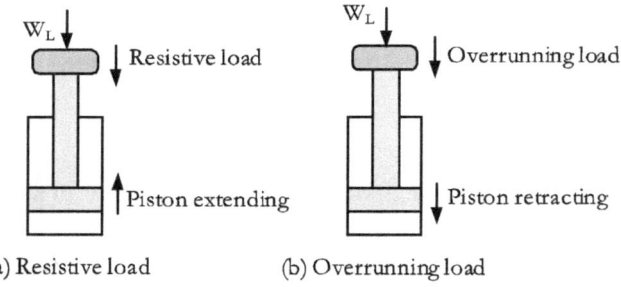

(a) Resistive load   (b) Overrunning load

Figure 2.3 | Types of loads associated with hydraulic systems

A resistive load on a hydraulic actuator tends to oppose the motion of the actuator, as shown in Figure 2.3(a). A cutting or shearing operation that acts against the motion of the cylinder is an example of a resistive load.

An overrunning load moves and acts in the same direction as that of the associated actuator. A descending heavy weight on a cylinder, as shown in Figure 2.3(b), is an example of an overrunning load.

An inertial load is a load that tends to resist the acceleration or deceleration of the actuator to which it is connected. A heavy flywheel or fan attached to the shaft of a motor is an example of an inertial load.

Summary of Relations for Hydraulic Cylinders

Figure 2.4 summarizes the essential relations of hydraulic cylinders, expressed in SI units for ease of understanding and to enable the reader to correlate these relations.

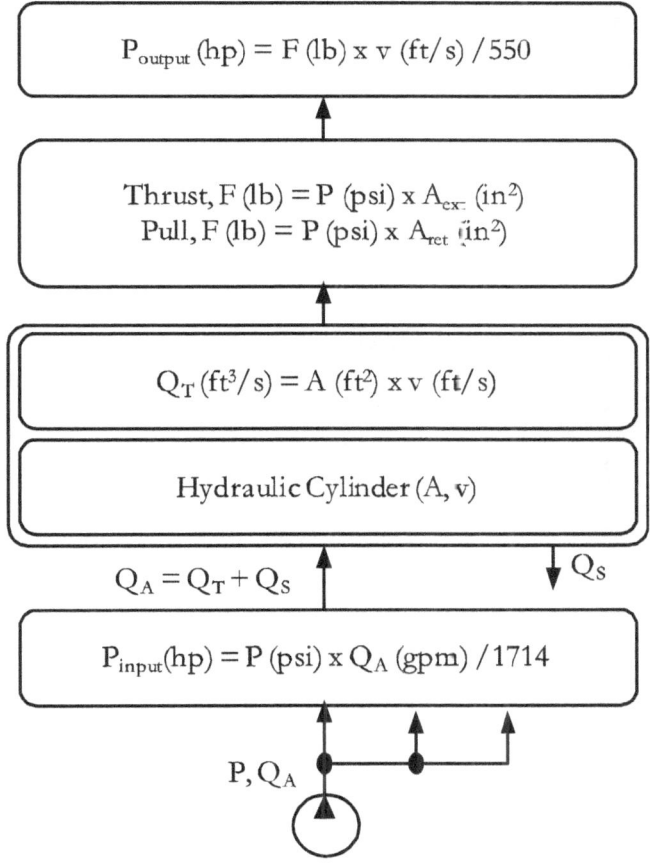

Figure 2.4 | Summary of relations - hydraulic cylinders

Example 2.2: Determine the thrust of a hydraulic cylinder with a bore diameter of 3.937 in, used for a flyover levelling application, and operating at a pressure of 10152 psi.

Solution

Bore diameter, D	= 3.937 in
Operating pressure, P	= 10152 psi

Piston area, extension stroke, $A = \pi D^2/4$
$$= \pi \times 3.94^2/4$$
$$= 12.19 \text{ in}^2$$

Thrust, F	= P x A
	= 10152 x 12.19
	= 123752 lb

Example 2.3: A double-acting hydraulic clamping cylinder must move outward at a velocity of 1.64 ft/s during the extension stroke. Calculate the flow rate required by the cylinder operating at 3000 psi and producing a thrust of 11240 lb.

Solution

Thrust, F	= 11240 lb
Pressure, P	= 3000 psi
Speed, v	= 1.64 ft/s
	= 19.68 in/s

Area, A_{ext}	= F/P
	= 11240/3000
	= 3.75 in²

Flow rate, Q	= A_{ext} x v
	= 3.75 x 19.68
	= 73.8 in³/s

Chapter 3 | Principal Parts and Body Styles of Hydraulic Cylinders

Hydraulic cylinders are designed to operate at high pressures and handle heavy loads under demanding conditions. Therefore, they need to be constructed from high-strength materials, with expert workmanship and advanced features, to provide ruggedness, high quality, maintenance-free operation, and a long service life.

A hydraulic cylinder consists of essential and optional parts, including a barrel, a piston, a piston rod, end caps, cushion seals, wear bands, a piston rod seal/wiper, piston rod bearings, piston rod boots, and a stop tube.

A hydraulic cylinder may incorporate advanced features, such as a magnetic piston and end-of-stroke or mid-stroke sensors, to measure piston position economically and reliably. Air bleeds can also be used to exhaust accumulated air from the cylinder.

Figure 3.1 shows the cross-sectional view of a simple double-acting hydraulic cylinder.

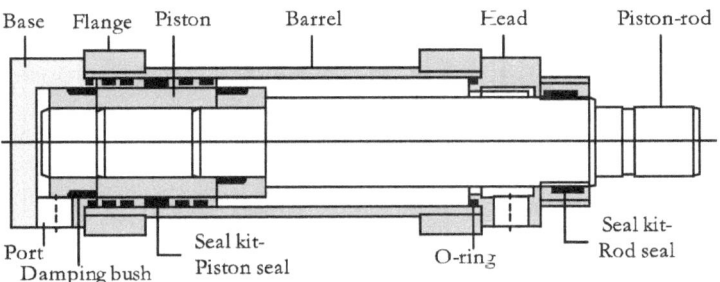

Figure 3.1 | A cross-sectional view of a hydraulic cylinder

It may be noted that a cylinder with leak-proof piston and piston-rod seals delivers consistent performance at very low speeds across a wide range of load resistances.

Barrel

The barrel of a hydraulic cylinder is made of a high-strength seamless-drawn tube, precision-machined to a perfect finish. The internal surface of the barrel must be highly smooth to minimize wear and leakage within the cylinder. A high-quality manufacturing process for the barrels ensures a smooth surface finish and the straightness and roundness of the cylinders, which, in turn, guarantee smooth action and superior fluid sealing properties. An aluminium, brass, or steel tube with a hard-chromed bearing surface (Figure 3.2) can be used as part of the barrel of a hydraulic cylinder if it is likely to be exposed to corrosive fluids. These materials are also better heat conductors, helping dissipate heat from applications with high-frequency cyclic operations. However, when using an aluminium barrel body, the cylinder should not be subjected to shock or high-inertia loading.

Figure 3.2 | Tubes for cylinder barrel

Piston

Figure 3.3 shows the schematic diagram of a piston. The primary function of the piston in a hydraulic cylinder is to transmit the force to the load attached to its piston rod. Additionally, it serves as the bearing in the cylinder barrel.

The piston must be a perfect fit inside the cylinder barrel. It must be reasonably cylindrical and finely finished for its smooth output motion. It is made of fine-grained alloy steel with a bronze coating. It can be threaded or welded to the piston rod. The grooves on the piston are provided to contain the packing or seals. For a heavy-duty application, an additional bearing ring may be fitted to the piston to improve wear resistance.

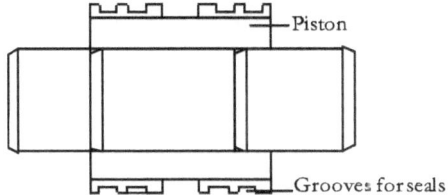

Figure 3.3 | Schematic diagram of a cylinder piston

Piston rod

The piston rod of a hydraulic cylinder moves in and out of the cylinder barrel and contacts the surrounding atmosphere. A smooth, hard, corrosion-resistant surface is essential for the piston rod's outer surface. Therefore, the piston rod is usually made of induction-hardened steel or stainless steel (Figure 3.4). It may also be chrome-plated with an ultra-fine surface finish to enhance wear and corrosion resistance. Good slide rings are provided in the piston rod for their excellent wear resistance and proper sealing. A piston rod may be of a hollow design or externally guided, or both.

Figure 3.4 | Piston rod

End caps

The end caps are shown in Figure 3.5. They are attached to the ends of the cylinder barrel to enclose the cylinder's pressure chamber. They are cast from iron or aluminium or made from high-quality steel. They may be designed with square or round shapes to match the barrel shape. They can be fixed with tie-rod ends or threaded or welded to the barrel. They also incorporate threaded entries for ports. They have to withstand any bending stress and shock loads to which they may be exposed. The end-

13

of-travel shocks in a cylinder can be absorbed by the cushion valves built into its end caps.

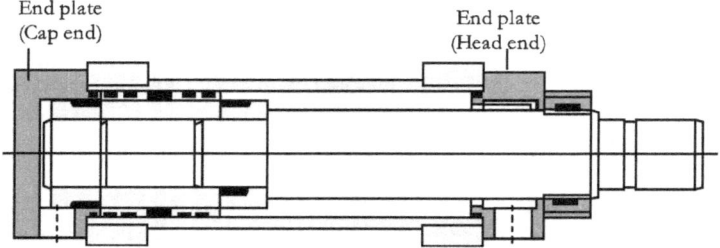

Figure 3.5 | End caps attached to the barrel

Cushion

It is an optional device in a hydraulic cylinder, incorporated at one or both ends, to minimize shock loads as the piston approaches its end-of-stroke position. A hydraulic cylinder should be equipped with cushions when the piston velocity is expected to exceed 4 inches/second. The details of cushion cylinders are given in a later chapter.

Seals

A small amount of leakage up to 3 in^3/min across the piston of a hydraulic cylinder is considered normal. Significant leakage can be a serious problem, leading to energy waste, reduced efficiency, environmental contamination, and safety hazards.

Leakage in a cylinder can be controlled by using appropriate sealing devices (Figure 3.6). A seal is an elastomeric part used in a hydraulic component to close the imperfections in their mating surfaces.

Installing proper seals in the cylinder helps maintain the system pressure, prevents fluid loss, reduces friction, and keeps contaminants out of the system. However, no sealing device is 100% reliable.

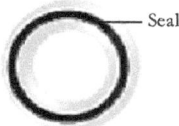

Figure 3.6 | Hydraulic seals

Sometimes a hydraulic cylinder seal may be deliberately designed to allow a controlled amount of internal leakage to lubricate its moving parts. A good seal must be compatible with a wide range of fluids and withstand the harsh industrial environment.

Piston Seals

Piston seals, as shown in Figure 3.7, are fitted into the grooves of the cylinder piston. Filled PTFE lip seals with internal stainless steel springs can be used as piston seals to optimize seal performance.

Pressure in the cylinder causes the cup seal lip to expand and grip the barrel, providing a positive seal. However, it can seal only in one direction. Therefore, double cup seals must be used in double-acting cylinders to provide sealing in both directions.

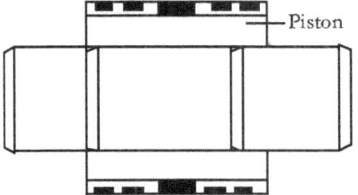

Figure 3.7 | Piston seals

Wear Bands

Wear rings/bands, as shown in Figure 3.8, in a hydraulic cylinder serve to limit the wear on its piston and piston rod seals, and provide superior protection against side loads. They are usually made from glass-reinforced nylon, which has excellent wear resistance and non-scoring properties.

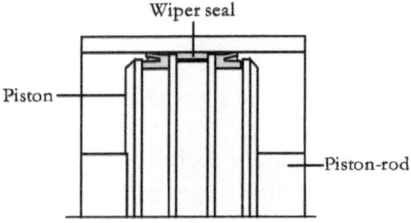

Figure 3.8 | Wear bands

Piston rod Seal/Wiper

Figure 3.9 shows the piston rod seal and wiper seal. These seals can be energized with fluorocarbon O-rings to maintain consistent contact pressure on the piston rod. The specially designed piston rod seal prevents the pressurized fluid in a hydraulic cylinder from leaking out while allowing the piston rod to move freely back and forth through the piston rod gland.

The wiper/scraper part of the seal prevents external contamination, such as dust and dirt, from entering the cylinder through the piston rod gland.

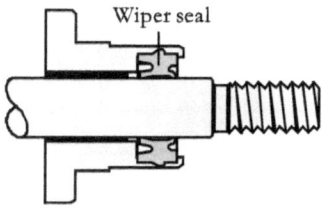

Figure 3.9 | Piston rod seal/Wiper

The seal also prevents damage to the bearing, seals, and piston rod. It consists of a honed wiping element made of age-hardened beryllium copper, surrounded by a synthetic rubber elastomer.

During the operation, the beryllium copper ring polishes the rod without damaging its surface finish.

Piston Rod Bearing

The piston rod bearing in a hydraulic cylinder guides the piston rod as it passes back and forth through the rod gland. It also supports the piston rod's weight and the load attached to it. Piston rod-end bearings are made of brass or bronze to handle side loads on the piston rod and ensure proper lubrication.

Piston Magnet

A cylinder piston can be equipped with a magnet to measure its position economically and reliably. By incorporating sensing capability into cylinders, the need for standalone proximity sensors can be eliminated.

Air Bleeds

Hydraulic cylinders are considered self-bleeding when cycled full stroke. However, air bleeds can be recessed into one or both ends of the cylinder body to vent trapped air.

For bleeding air, hydraulic pressure is typically set between 150 and 400 psi. High pressure should not be used during bleeding to prevent fluid spray injury. First, move the piston to the end-of-stroke position, and then slightly open the bleeder screw until bubble-free oil emerges. Then tighten the bleeder screw again. Bleed the other side of the cylinder in the same way.

Piston Rod Boots

When dirt, chips, scale, and other contaminants contact exposed cylinder rods, they not only shorten the seals' service life but also damage the rods. In highly contaminated environments, the exposed piston rod should be covered with a boot or bellows (Figure 3.10). The piston rod boot is a collapsible neoprene-coated fabric cover that is water- and oil-resistant. This arrangement protects the piston rod, seals, and bearing from contamination.

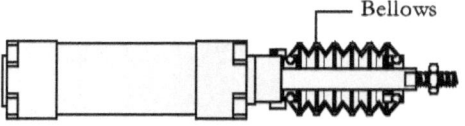

Figure 3.10 | Piston rod boots

Stop tube

The stop tube is used in a horizontally mounted hydraulic cylinder, especially with a long-stroke, poorly guided piston rod.

Figure 3.11 shows the schematic diagram of the stop tube. It is a separator between the head-end cap and the piston of the cylinder, intended to prevent the piston from reaching the end of the cylinder barrel. This separation reduces the moment forces acting on the piston rod bearing when the rod is fully extended. Reducing the moment forces helps maintain bearing loads within acceptable limits and increases the cylinder's side-load capacity and stability compared to a standard cylinder without a stop tube.

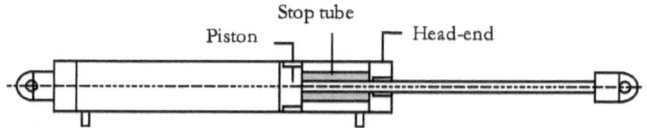

Figure 3.11 | A hydraulic cylinder with a stop tube

Body Styles of Hydraulic Cylinders

Hydraulic cylinders are manufactured in a variety of body styles based on how their outer parts are assembled. Based on their assembly type, hydraulic cylinders fall into four basic categories.

They are: (1) Tie-rod cylinders, (2) Mill type cylinders, (3) Threaded-end cylinders, and (4) Welded cylinders.

Any of these body styles can be selected based on the application and the surrounding environment.

Tie-rod Hydraulic Cylinders

This type of cylinder is most commonly used in industrial hydraulic applications. In a tie-rod cylinder, four or more high-strength threaded steel tie rods run the length of the cylinder. The tie-rod assemblies securely hold the barrel and two end caps together. The size and strength of the tie-rods and their fasteners determine the cylinder's strength. High-tensile tie rods in a cylinder commonly bear a large portion of the applied load and absorb internal stresses. A large-bore high-pressure tie-rod cylinder may have as many as 24 tie rods. Figure 3.12 shows the front and end views of a tie-rod cylinder.

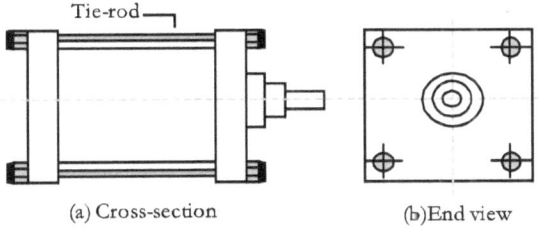

| (a) Cross-section | (b) End view |

Figure 3.12 | Front and end views of a tie-rod cylinder

The tie-rod cylinders offer a rugged design, low maintenance, and a long service life. They can be fully disassembled for service and repair. Therefore, light- to medium-duty cylinders used in the automotive, mobile, and machine tool industries are of the tie-rod design. These sturdy, dependable, versatile cylinders can also be used in various hydraulic applications, including industrial manufacturing and material handling.

However, long-stroke tie-rod cylinders pose several problems, such as tie-rod stretch, tie-rod sag, and cylinder flexing and twisting. The long steel tie-rods can stretch to a certain degree under severe tension, and they may stretch to such an extent that the barrel separates from the end caps. A long steel tie rod also sags under its own weight. Therefore, it must be fitted with one or more intermediate supports along the barrel to prevent sagging. The tie-rod cylinders are also susceptible to twisting and flexing. Manufacturers publish pressure limits and mounting-style restrictions for cylinders due to the tie-rod's inherent stretch, flex, and twist.

Mill Type Hydraulic Cylinders

They are heavy-duty cylinders designed for harsh environments and demanding service conditions, as in steel mills, mines, and furnaces. A mill-type cylinder typically has a thick-walled tubular housing that provides exceptional strength and heat resistance. A cylinder of this type has its end caps fastened to the mating flanges with bolts or cap screws, as shown in Figure 3.13.

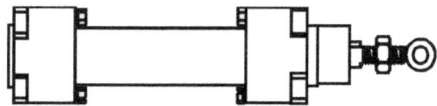

Figure 3.13 | Mill-type hydraulic cylinder

Mill-type cylinders are designed to resist buckling under compressive loads and can be used in both corrosive and non-corrosive environments. In addition to steel mills, furnaces, and mines, they are well-suited for other applications, including paper mills, oil and gas drilling rigs, metal-forming industries, and heavy construction.

Welded Hydraulic Cylinders

Figure 3.14 shows a welded hydraulic cylinder. The heavy-duty barrel of this type of cylinder is welded along its entire outer diameter to each end cap. Also, its ports are welded to the barrel, and the mountings are welded to the cylinder body. The welded cylinders provide an incredibly strong bond and high strength. They are also very compact. Moreover, they have smooth exterior surfaces, free of tie-rods and fasteners. These constructional features mean that there are fewer places for dirt and debris to accumulate. The round body is free of obstructions, making it easy to clean.

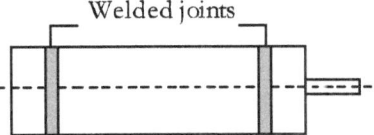

Figure 3.14 | A welded hydraulic cylinder

The welded cylinders are compact and usually designed for high strength. They are space-efficient in their overall length compared to tie-rod cylinders. They are also not susceptible to tie-rod sag or stretch. Because of these advantages, they dominate the mobile hydraulic equipment market and the heavy equipment industry. However, welded cylinders cannot be disassembled for servicing. In the event of a seal failure in a welded cylinder, repairs typically require specialized, heavy-duty equipment to cut and subsequently re-weld the cylinder.

Threaded-end Hydraulic Cylinders

In this type of cylinder construction, the cylinder end caps are threaded to match the internal threads of the tubular cylinder bore. In applications where space is a consideration, these cylinders can be used. The food-processing and food-packaging industries extensively use this cylinder body style.

Application Summary: Tie-rod cylinders are optimal for medium-pressure applications up to 3000 psi, including industrial machinery and factory automation. Mill-duty cylinders are suited to extreme conditions, including steel mills, marine environments, and heavy stamping presses. Welded cylinders are suitable for heavy-duty mobile equipment in the construction, agriculture, and mining sectors, especially in space-constrained environments. Threaded-end cylinders are ideal for medium-pressure applications that require a smaller profile than tie-rod cylinders and offer easier servicing than fully welded cylinders.

Note: Appendix 4 gives different types of piston seals, ports, and mounting styles. Appendix 5 gives seal materials and their temperature ranges.

Chapter 4 | Side Loads in Hydraulic Cylinders

The side load in a cylinder is the force component that acts laterally across the axis of its bearings, piston rod seals, and piston. A long-stroke hydraulic cylinder often experiences significant side loads during operation, especially when fully extended.

Many factors contribute to the development of side loads in a cylinder. An off-center load on the cylinder, as shown in Figure 4.1, causes side loads on its bearings, seals, and piston. Similarly, the cylinder experiences lateral loads when misaligned or improperly mounted. Even the weight of the cylinder or its extended piston rod exerts a side load on the rod bearings and piston bearings.

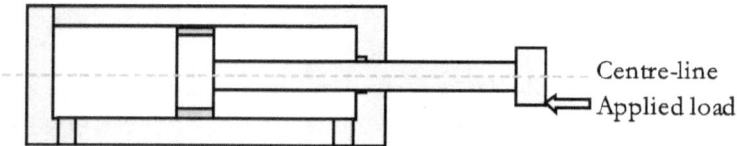

Figure 4.1 | A hydraulic cylinder applied with an off-centre load.

Side loads on a hydraulic cylinder can cause stress in its bearings, seals, piston, mounting flange, and mounting bolts. Lateral loads exert pressure on the piston rod on one side of the gland, causing seal damage and leakage. Accelerated, uneven wear occurs on the piston rod bearing and gland. They ultimately reduce the cylinder's service life.

It is essential to reduce the side loads imposed on long-stroke cylinders. It can be reduced by proper cylinder mounting. The cylinder axis must be aligned with the load axis.

The side load capacity of a cylinder can be improved by providing adequate bearing support, a heavy-duty wear ring, an oversized piston rod, dual pistons, and/or a stop tube.

Chapter 5 | Piston Rod Buckling

Cylinders capable of giving very long strokes are essential for some hydraulic applications. However, if a compressive axial load is to be applied to the piston rod of such a long-stroke hydraulic cylinder, it must be within the safety limit to prevent buckling of the piston rod.

Piston rod buckling occurs when the piston rod bends under load. The size of the piston rod of a long-stroke cylinder must be adequate to withstand the buckling stress. It depends on the cylinder mounting style and rod-end connection. A stop tube is recommended for horizontally mounted long-stroke cylinders to reduce bearing loads. Piston rod size can be found from the charts provided by cylinder manufacturers A chart from the catalogue 'Hydraulic and Pneumatic Cylinder – Application Engineering Data', Cylinder Division, Parker, is given in Figure 5.1.

Selection of Piston Rod for Push Stroke

The selection of the piston rod is as follows:

- Find the stroke factor from Table 5.1, as per the mounting style and rod-end connection.

- Determine the 'basic length' from the following equation, using the stroke factor.

Basic length = Actual stroke x Stroke factor

- Find the load imposed for the thrust application (F=PxA)

- Find the size of the piston rod from Figure 5.1 at the intersection point of the basic length and the thrust force.

- Also, find the length of the stop tube from the chart.

Table 5.1 | Stroke factors

Mounting style	Rod-end connection	Case	Stroke factor
Straight-line force transfer	Fixed and rigidly guided		0.5
Straight-line force transfer	Pivoted and rigidly guided		0.7
Straight-line force transfer	Supported, but not rigidly guided		2.0
Trunnion on the head	Pivoted and rigidly guided		1.0
Intermediate trunnion	Pivoted and rigidly guided		1.5
Trunnion (or clevis) on cap	Pivoted and rigidly guided		2.0

Piston Rod Selection Chart

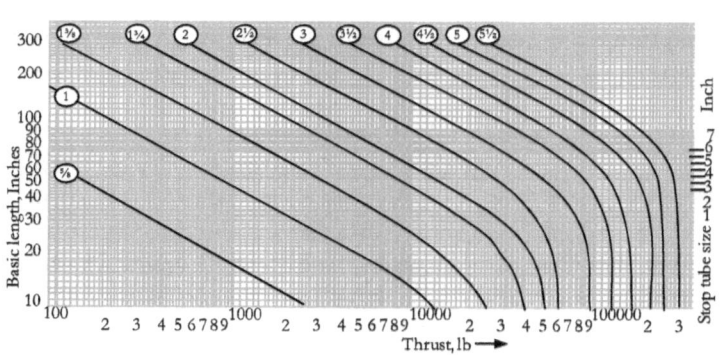

Figure 5.1 | Piston rod selection chart
Courtesy: Parker Cylinder Division

Chapter 6 | Classification and Types of Hydraulic Actuators

Hydraulic actuators are available in a wide range of types, sizes, and models to meet diverse application requirements. Table 6.1 provides a broad classification of hydraulic actuators, outlining their main types and subtypes.

Table 6.1 Classification of hydraulic cylinders

Main types	Sub-types	Examples
Linear actuators	Single-acting	Spring-returned
		Gravity returned
	Double-acting	Non-cushioned type
		Cushioned type
	Variants	Plunger/Ram cylinder
		Position-sensing cylinder
		Double-rod-end cylinders
	Special assemblies	Tandem cylinder
		Telescopic cylinder

Single-acting Hydraulic Cylinders

Figure 6.1 shows the cross-sectional view of a single-acting cylinder. It consists of a barrel, a piston-and-rod assembly, a spring, end caps, seals, and a port. A fluid chamber is formed in the cylinder with the barrel, piston, and cap-side endplate. The piston-and-rod assembly is a tight fit inside the barrel and is biased by the spring. The seals are primarily used to prevent fluid leakage in the system. The port is integrated into the cap end for applying or relieving the system fluid. Applying hydraulic pressure through the port moves the piston-and-rod assembly in one direction, providing the working stroke. The piston-and-rod assembly moves in the opposite direction under spring force, gravity, or an external force. In a cylinder with a spring-assisted retraction, the spring is designed not to carry any load, but to retract the piston and piston rod assembly with sufficient speed.

In an alternative design of a single-acting cylinder, the spring can be mounted on the piston side, with the port at the head end. The single-acting cylinder is capable of performing work only in one direction of its motion, and hence the name 'single-acting cylinder'.

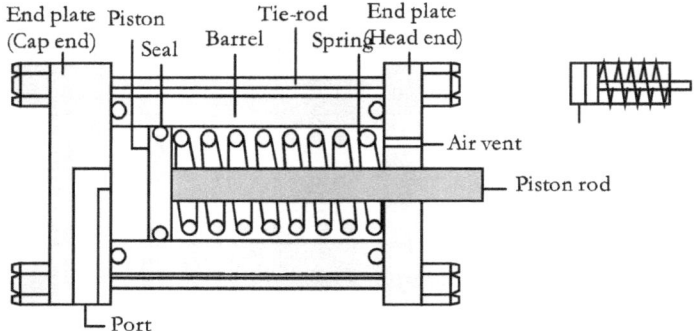

Figure 6.1 | A cross-sectional view of a single-acting cylinder

The cylinder's stroke length is typically limited to 4 inches due to the spring's natural length. Single-acting cylinders are simple, economical linear actuators. A single-acting cylinder converts system pressure and flow into mechanical force and motion, respectively. The ease of operation of the single-acting cylinders makes them particularly suitable for applications such as clamping, pressing, cutting, holding, ejecting, feeding, and lifting.

Double-acting Hydraulic Cylinders

Figure 6.2 gives the cross-sectional view of a double-acting cylinder. It consists of a barrel, a piston-and-rod assembly, end caps, seals, and two ports. In contrast to the single-acting cylinder, the double-acting cylinder has fluid ports at both ends: the piston-side port and the piston-rod side port. Applying pressure through the piston-side port extends the cylinder, provided that pressure on the piston-rod side is relieved. Similarly, applying pressure to the piston-rod side port retracts the cylinder, provided that pressure on the piston side is relieved. That means the cylinder can convert pressure and flow into mechanical force and motion, respectively. A double-acting cylinder can provide power strokes in both directions of its motion, hence the name 'double-acting cylinder'.

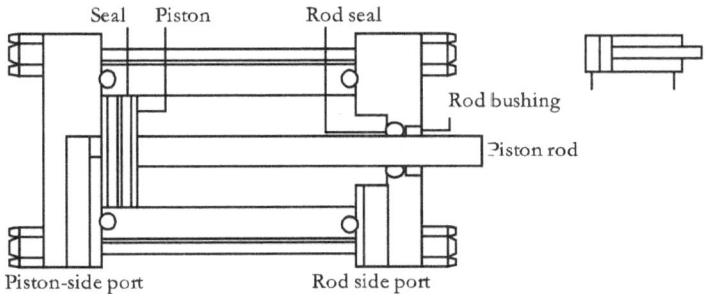

Figure 6.2 | A cross-sectional view of a double-acting cylinder

Ideally, a conventional double-acting hydraulic cylinder can be designed to have an unlimited stroke length. However, the practical maximum stroke length of the cylinder is limited to about 6.5 ft due to potential bending and buckling of the extended cylinder with a very long piston rod.

Double-acting hydraulic cylinders are the most common type in modern industry. They can be used in applications that require mechanical power for linear motion. They provide high push and pull forces for the connected loads. They perform well in applications, especially where precise low-speed control is required. They are designed for heavy-duty lifting and positioning in mobile applications and in industrial production and assembly.

Hydraulic Cylinders - Differential Vs Non-differential

In a typical single-ended double-acting hydraulic cylinder, as shown in Figure 6.3(a), the piston area (say A1) exposed to fluid contact on the piston end is larger than that (say A2) on the piston rod end. The active area of the fluid contact on the piston rod end is equal to the bore area minus the cross-sectional area of the piston rod. Such a cylinder is regarded as a differential cylinder.

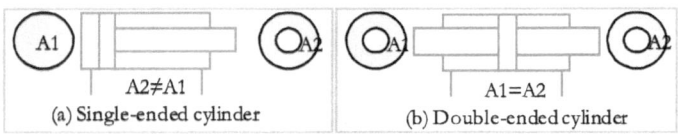

A2≠A1	A1=A2
(a) Single-ended cylinder	(b) Double-ended cylinder

Figure 6.3 | Basic types of hydraulic cylinders

The differential cylinder produces a slightly greater force and lower speed during extension than during retraction, provided the pressures at both ends of the cylinder are equal. The required area ratio (A1/A2) for a differential cylinder depends on its construction, application, and stroke length.

For standard applications, the area ratio is typically about 6:5. For heavy-duty applications with large-size cylinders and piston rods, an area ratio of about 2:1½ may be selected.

In a non-differential hydraulic cylinder, as shown in Figure 6.3(a), the areas of fluid contact on the piston end and the piston rod end are the same. This fact enables equal forces and speeds in both directions of piston travel.

Cushioning in Hydraulic Cylinders
The load-attached fast-moving piston of an ordinary hydraulic cylinder produces impact forces when it strikes its end covers. These end-of-travel shock loads can be reduced by incorporating cushioning devices in the cylinder, either at one end or at both ends.

The cushioning devices decelerate the piston-and-rod assembly as it approaches its end-of-stroke position, thereby preventing excessive mechanical stresses. The cushions are either fixed or adjustable. Hydraulic cylinders can be designed to operate at higher speeds with simple or adjustable cushions.

Hydraulic Cushion Cylinder

Figure 6.4 shows the cross-sectional view of a double-acting hydraulic cylinder with adjustable cushioning devices. A cushioning device comprises a throttle valve and a check valve integrated into the cylinder end caps. Cushion sleeves are attached to the piston.

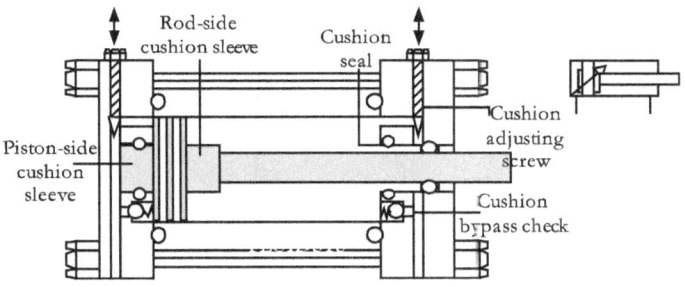

Figure 6.4 | An adjustable cushion hydraulic cylinder

When the piston moves forward, the full system flow can exit unrestricted through the large orifice in the fluid chamber at the head end of the cylinder. As the rod-side cushion sleeve enters its chamber, the system fluid's normal exit path is blocked. This obstruction forces the flow through the throttle valve, which restricts it and progressively slows the piston's movement.

The piston continues to move to its end-of-stroke position at a controlled speed without damaging the cylinder. The cushions are usually designed to operate over the final inch of the piston stroke.

In many cylinders, the cushioning can be adjusted by using an adjusting screw. Typically, the cap-end cushion would function in the same way as the head-end cushion.

A bypass check valve can be incorporated into the cushioning device to allow the unrestricted entry of the system fluid into the cylinder.

Some pressure intensification occurs at the piston rod end, but it is usually not a significant problem.

If the load-coupled piston tends to generate high forces and accelerations, additional measures, such as installing an external shock absorber, must be implemented to assist in load deceleration.

Ram (Plunger) Cylinders

A variant of the standard single-acting hydraulic cylinder is the ram (plunger) cylinder. Figure 6.5 shows the cross-sectional view of the ram cylinder.

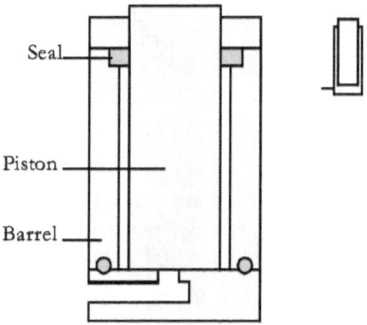

Figure 6.5 | A Plunger/Ram cylinder

A ram cylinder is designed with a robust and thick piston rod (ram) whose diameter is almost equal to that of the cylinder piston. Most ram cylinders do not use return springs. Instead, they use gravity or the loads attached to them to retract the piston rods.

Both the piston and the piston rod in a ram cylinder must be precisely cylindrical and finely finished for the smooth output motion. A surface-hardened or chrome-plated piston rod is often used with an ultra-fine surface finish to ensure the long service life of its sealing elements.

The ram cylinder is typically mounted upright and used for short-stroke applications. The ram cylinders are primarily used for pushing rather than pulling. However, a ram cylinder with a hollow rod can be used for both pushing and pulling operations. Piston rod bending in a basic long-stroke horizontal cylinder unit, or buckling of the basic cylinder under a high vertical load, can be prevented by replacing the basic cylinder with a ram cylinder.

The ram cylinders are most commonly used in high-pressure applications, such as pressing and stamping operations in machines and lifts in automobile service stations.

Double-rod-end Cylinders
A variant of the standard double-acting cylinder is the double-rod-end cylinder. Figure 6.6 shows the double-rod-end cylinder, with piston rods extending from both ends.

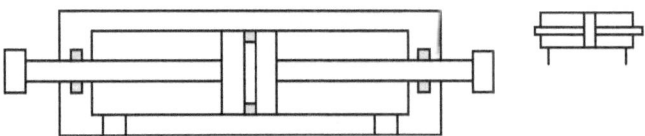

Figure 6.6 | A double-rod-end hydraulic cylinder

The double-rod-end arrangement provides better rod alignment because the attached piston rods move on two bearings.

The cylinder has equal areas on both sides of the piston and operates without the differential cylinder effect. Using a double-rod-end cylinder, two functions can be performed at either end in staggered cycles. It is also used when the piston speed must be the same in both directions of its motion.

Furthermore, it is useful in a robotic mechanism in which the piston rods are clamped at both ends of the cylinder, and the body moves instead.

Telescopic Cylinders

A multi-stage telescopic cylinder nests several cylinder bodies inside one another. That is, the piston rod of the first stage is used as the piston barrel of the second stage, and the second piston rod is used inside the barrel of the second stage. Similarly, there can be up to six stages. Therefore, the total stroke length of the telescopic cylinder can be up to six times the stroke length of the primary cylinder. Remember that the output force is highest in the first stage and decreases with each subsequent stage.

The telescopic cylinders are ideal for applications requiring long-stroke cylinders in space-constrained environments. They are widely used in hydraulic equipment for the agriculture, construction, and heavy engineering industries. They are commonly used in mobile hydraulic systems for the tilting of truck dump bodies and forklifts, lifting operations in hydraulic cranes, and material handling. The telescopic cylinders are available in single- and double-acting configurations.

Single-acting Telescopic Cylinder

Figure 6.7 shows the single-acting telescopic cylinder. It is a multi-stage cylinder with two to six concentric tubular envelopes. Most telescopic cylinders are single-acting, meaning the fluid pressure always acts in one direction. That is, on the forward strokes, to be more precise. Single-acting telescopic hydraulic cylinders are used for heavy-duty, one-way lifting operations. These applications include dump trucks and agricultural machinery.

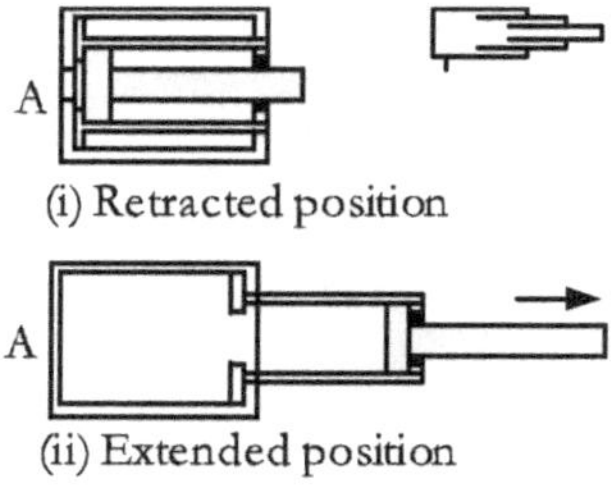

(i) Retracted position

(ii) Extended position

Figure 6.7 | A single-acting telescopic hydraulic cylinder

Double-acting Telescopic Cylinder

Figure 6.8 shows the double-acting telescopic cylinder. In this type of cylinder, the system pressure alternately extends and retracts the cylinder. The double-acting telescopic cylinders are highly complicated and must be specially designed and manufactured to a high degree of precision Therefore, they are much more expensive than conventional hydraulic cylinders.

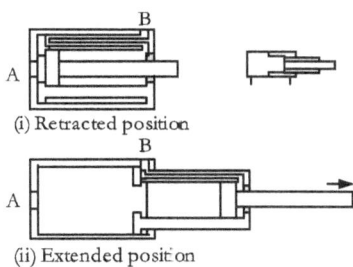

(i) Retracted position

(ii) Extended position

Figure 6.8 | A double-acting telescopic hydraulic cylinder

Tandem Cylinder

In the tandem type of cylinder, two (or more) cylinders are assembled in-line, with the piston rod of the first cylinder attached to the piston of the second cylinder, and so on. Figure 6.9 shows the cross-sectional view of the tandem cylinder.

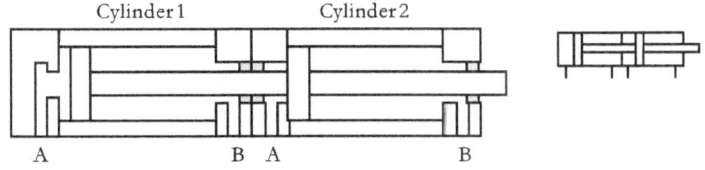

Figure 6.9 | A cross-sectional view of a tandem cylinder

When pressurised fluid is applied to the piston-side ports (marked as 'A'), the cylinder extends with force, which is twice as much as the force produced by a single-stage cylinder of the same bore size while extending. In the same way, when pressurised fluid is applied to the piston rod-side ports (marked as B), the cylinder retracts with force, which is twice as much as the force produced by a single-stage cylinder of the same piston and piston rod sizes while retracting.

The tandem cylinder is suitable for applications where a high output force must be developed within a narrow radial space, but there is sufficient axial length. Key applications of tandem cylinders include industrial machinery, construction equipment, and material handling.

Swing Clamp Hydraulic Cylinder

It is a specialized cylinder with a clamp arm. It can swing and clamp. The piston and piston rod assembly of the cylinder can rotate by a certain angle clockwise or counterclockwise during the swing stroke, and then travel in a straight line during the clamp stroke. It is designed for secure, obstruction-free clamping and unclamping of workpieces. They are designed to operate at approximately 1000 psi, providing significant forces.

However, the arm should not contact the workpiece during the swing stroke. Swing cylinders are available in single-acting and double-acting versions, with an integrated return spring. In applications where return time is critical, a double-acting cylinder can ensure positive retraction within the required timeframe.

Construction of Swing Cylinder

Figure 6.10 shows the cross-sectional views of a swing cylinder in the extended and retracted positions. It consists of a barrel, a piston, and a piston rod with an arm or lever. The piston is a tight fit in the barrel. A standard-length arm is attached to the piston rod on one side of the piston for clamping. The piston rod on the other side of the piston has ball seats with straight and angled sections along their axes, as shown. Cam followers made of tungsten carbide contact and guide the ball seat on the piston rod. This sturdy swing mechanism controls the piston's swing and straight motions during extension and retraction.

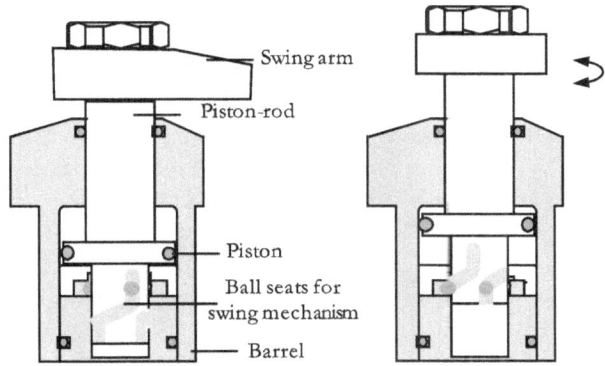

Figure 6.10 | Cross-sectional views of a swing clamp cylinder in the extended and retracted positions

Working Principle of a Swing Cylinder

The swing cylinder clamps the workpiece during retraction and unclamps during extension. The movement of the piston and piston rod assembly with an arm consists of a swing stroke and a clamp (or unclamp) stroke. The piston-and-piston-rod assembly with the arm turns during the swing stroke. It then retracts in a straight line to apply clamping force, with the arm in position over the workpiece during the clamping stroke. It extends straight as it releases the workpiece during the unclamp stroke. Then the piston, along with the arm, rotates to the fully extended position during the swing stroke, providing easy access for loading and unloading the workpieces.

Typical Specifications, Swing Cylinders

Table 6.1 | Specifications of swing cylinders

Bore size	½ to 3 inches
Clamping stroke	¼ to 2 inches
Swing angle	0°, 30°, 45°, 60°, 90°
Maximum operating pressures	7250 psi

Chapter 7 | Position Transducers for Hydraulic Cylinders

As demand for greater functionality and easier control of hydraulic cylinders increases in heavy industry and mobile equipment, the need for hydraulic cylinders equipped with position transducers (sensors) is becoming more critical. A position transducer is essentially a feedback device used with a cylinder that senses the position of the cylinder's piston rod.

A position transducer typically consists of two parts: one fixed to the cylinder body, and the other moving with the piston rod whose position is being measured. It can be mounted externally or integrally to the piston rod. If a transducer is to be mounted inside a cylinder, a hole must be drilled through the piston rod. Further, the cylinder end cap is machined to accommodate the transducer. The initial cost of gun-drilling the rod and the replacement cost can limit the use of integrally mounted transducers.

Three types of position transducers are typically used in hydraulic cylinders. They are: (1) Variable resistance potentiometers, (2) Linear Variable Differential Transducers (LVDTs), and (3) Magnetostrictive transducers.

Linear Resistance Potentiometers

A linear resistive transducer is a linear potentiometer used to measure piston rod position. It can be an external cable connected to a rotary potentiometer or an internally mounted flat potentiometer contained within the piston rod. The piston rod supports an electrically conductive wiper running along the surface of a partially conductive plastic probe. As the wiper moves over the plastic element, its resistance varies linearly, making it easy to determine the piston rod position. Linear resistive transducers are simple, robust, capable of handling high shock and vibration, and low-cost devices, but they have low accuracy. They also wear out over time.

Linear Variable Differential Transducers (LVDTs)

An LVDT is a non-contact transducer that converts linear displacement into an electrical output signal. It consists of primary and secondary coils. This coil assembly is then installed in the cylinder piston. A ferromagnetic core moves inside the coils. An alternating current in the primary winding creates an axial magnetic flux. This flux is coupled to the secondary windings through the core, inducing an output voltage in each. The series-connected secondary windings produce a net voltage relative to the core position, using built-in electronics.

LVDTs have the highest accuracy and can withstand shocks and vibrations in the associated systems. They supply 4-20 mA or 0-10 VDC analog signals for closed-loop control.

Magnetostrictive Transducers

Magnetostriction is a property of ferromagnetic materials that causes them to change their shape or dimensions in the presence of a magnetic field.

A magnetostrictive linear position transducer, as shown in Figure 7.1, uses an iron-alloy sensing element, typically referred to as a waveguide. The waveguide is housed inside a pressure-rated stainless steel tube.

Further, a conductor and signal converter are attached to the waveguide. The piston of a hydraulic cylinder is fixed with a permanent magnet.

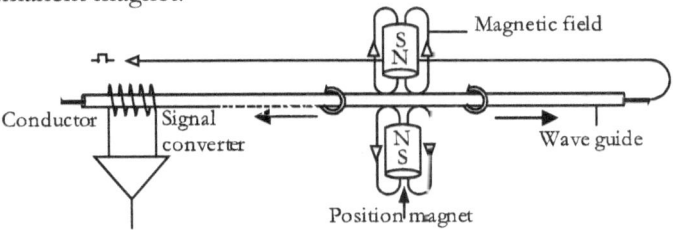

Figure 7.1 | A magnetostrictive transducer

Initially, a short-duration electrical pulse (1 − 3 μsec) is applied to the conductor. The current creates a magnetic field along the waveguide.

This magnetic field interacts with the field generated by the position magnet, inducing a torsional deflection of the waveguide element.

When the current pulse ends, the strain is abruptly relaxed, propagating a mechanical pulse along the waveguide. This mechanical pulse travels in both directions at a constant speed (9350 ft/s) and is detected at the signal converter.

Detection of the mechanical wave in the signal converter completes one measurement cycle.

The time between the initial electrical pulse and the received mechanical pulse accurately represents the magnet's absolute position and, hence, the hydraulic cylinder's position.

The magnet's position along the waveguide is determined by measuring the time interval between the initial current pulse and the detection of the mechanical return pulse.

Measurement cycles are typically repeated at 0.5-5 ms intervals, depending on the sensor length.

Linear position sensors are the ideal choice for hydraulic cylinder position feedback. They are available in measuring lengths from 2" to 300" and can withstand high-pressure operation up to 8700 psi.

Many electrical interface options are available to adapt to any control system.

They are exceptionally well-suited for hydraulic cylinders in industries such as steel processing and tire manufacturing.

Comparison of Position Transducers

Selecting a position transducer/sensor for hydraulic cylinders requires balancing cost, durability, and precision. A comparison of position transducers is given in Table 7.1.

Table 7.1 | Comparison of position transducers for hydraulic cylinders

Feature	Linear Potentiometer	LVDT (Inductive)	Magnetostrictive (Magnetic)
Technology	Contacting (Wiper)	Non-contacting (Inductive)	Non-contacting (Magnetic)
Ruggedness	Good	Excellent	High
Accuracy and resolution	Moderate to Low	High	Excellent (Micron level)
Cost	Cheapest	Moderate	High cost
Life Expectancy	Limited due to wear	Extremely Long	High as there are no wearing parts
Drawback	Susceptible to wear	Sensitive to external magnetic field	More expensive, lower temperature tolerances
Applications	Suitable for cost-sensitive applications	Suitable for harsh environments, high-cycle applications	Suitable for high precision, long stroke, high-pressure applications

Chapter 8 | Installation and Mounting of Hydraulic Cylinders

The hydraulic cylinder body must be mounted properly, and its piston rod must be coupled to the machine for optimal cylinder performance. Any mismatch in the mounting or coupling can lead to side loads. Remember, the side loads act laterally across the barrel, piston, piston rod, seals, and bearing bushes. Side loads will cause stresses in the cylinder components, reducing the cylinder's service life.

A hydraulic cylinder should be so fixed that the side loads on the piston rod bearing must be a minimum. By adhering to sound engineering practices, they can be reduced to an acceptable level. Slide or roller guides can be used to carry the load, wherever possible. The mounting dimensions of hydraulic cylinders must comply with the relevant ISO/NFPA standards.

Mounting Methods of Hydraulic Cylinders

A cylinder can be mounted to a machine using a permanent or movable mounting. For example, the body and piston rod of a hydraulic cylinder can be rigidly fastened to the machine (fixed-type mounting), or the cylinder can be allowed to swivel as part of the linkage in one or more planes (pivot-type mounting).

Figure 8.1 shows the schematic diagrams of these two configurations.

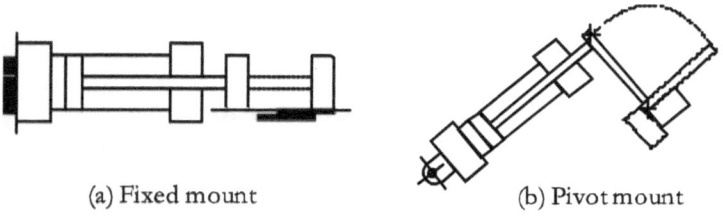

(a) Fixed mount   (b) Pivot mount

Figure 8.1 | Mounting arrangements for hydraulic cylinders

The choice of mounting method for a hydraulic cylinder depends primarily on the application's characteristics. Factors such as cylinder stroke length, piston rod diameter, method of load connection, and the presence of shock pressures must be considered when selecting the cylinder mounting style.

Mounting Styles of Hydraulic Cylinders
Cylinder mounts and accessories are available in various configurations. The mounts can be fixed or movable. Typical examples of fixed-mountings include tie-rod, flange, and foot mounts; typical examples of pivot-mounts include head, cap, or center trunnions, swivel flanges, and rear cap pivot mounts. The load-bearing capacity of a cylinder mount should exceed the load of the associated cylinder. Some critical mounting methods for the cylinder body are depicted in Figure 8.2(a), and some coupling methods for the piston rod are depicted in Figure 8.2(b).

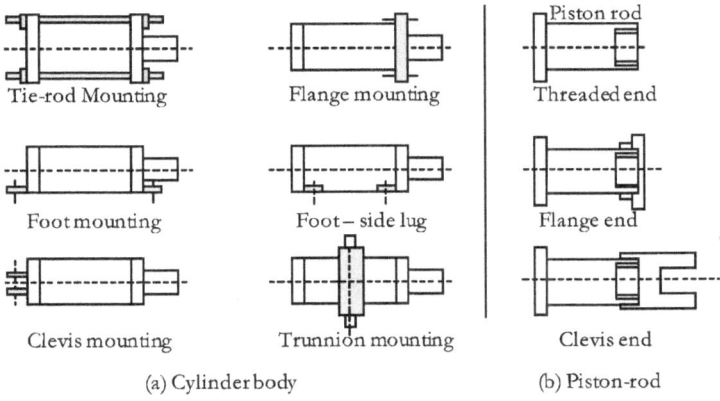

Figure 8.2 | Typical mounting arrangements of cylinders

Tie-rod mount: In this type, the extended tie rods of a hydraulic cylinder are used to mount the cylinder. The tie-rod can be extended on either the piston side or the piston-rod side. The free end of the cylinder should be supported to prevent its misalignment or sag. The best application of the tie-rods extended on the piston side is with a thrust load (i.e., with the piston rod in compression), and that of the tie-rods extended on the piston rod side is with a tension load.

Flange Mount: This type of mount provides a sturdy and rigid mounting for a cylinder. With this type of mount, there is no scope for misalignment. The flange mounting is best for installing the cylinder when the load acts along the cylinder's centerline. The best use of a cap-end flange in a cylinder is for thrust loads. The piston rod-end flange mounts are best suited for tension loads.

Foot or Lug Mount: The foot-mounted or lug-mounted cylinder provides relatively rigid mounting. This type of cylinder mounting can tolerate some amount of misalignment.

Pin-and-Trunnion Mounts: A pin-and-trunnion-mounted hydraulic cylinder requires provisions for pivoting at both ends. It is designed to carry shear loads. If the load path is curved or misalignment is a concern, the pivoted centerline mounting should be used.

Piston rod Mounts: Piston rod coupling accessories include pinhole, rod-end threads, rod clevis, eye bracket, knuckle, and pivot pin. Figure 8.2(b) shows some coupling arrangements for the piston rods.

Threads: Cylinder threads are essential for ensuring leak-proof operation, maintaining structural integrity, and facilitating proper mounting. They are generally classified into port threads, which allow oil entry, and rod or structural threads, which are designed for mechanical connections. The most commonly used threads are the Metric Thread, British Standard Pipe Thread (BSPP), National Pipe Thread (NPT), and Unified Fine Thread (UNF). Essential specifications for the piston rod end include its thread diameter and pitch, and the thread length. [Also see Appendix 6.]

Chapter 9 | Advantages of Hydraulic Cylinders

Some of the essential advantages of hydraulic cylinders are their ruggedness, power density, and economy. They are also available in a broad range of strokes. The following section explains these positive features:

Power Density: A hydraulic cylinder can deliver a high force-to-weight ratio for its relatively small size. This powerful actuator can be built to fit the tight confines of modern machinery. Hydraulic cylinders can deliver high force to move heavy loads in industries such as mining and construction.

Ruggedness: Hydraulic cylinders are rugged actuators. If a hydraulic cylinder is highly overloaded, it simply stops. It does not overheat.

Reliability and Safety: Hydraulic cylinders are highly reliable, even in harsh industrial environments. They are robust, often include built-in safety mechanisms against overloading, and, due to fewer moving parts, require less maintenance. Built with rugged materials, they operate reliably in high-heat conditions.

Range of Strokes: Hydraulic cylinders are available with a wide range of strokes. Small industrial hydraulic cylinders, such as clamping cylinders, may have a maximum stroke of ¼ inch. Large hydraulic cylinders used in the heavy equipment industry can have power strokes up to 6.5 ft. Telescopic cylinders can achieve very long stroke lengths.

Economy: Hydraulic cylinders are relatively easy to manufacture to the exact dimensions and speed requirements specified by end users.

Versatility: They can be customized for specific stroke lengths, mounting configurations, and bore sizes, making them adaptable to a wide range of applications.

Chapter 10 | Applications of Hydraulic Cylinders

Manufacturers offer a range of standard cylinders and heavy-duty cylinders suitable for a variety of applications and harsh environments.

Hydraulic cylinders can be single-acting or double-acting, telescopic, or plunger-type.

The heavy-duty hydraulic cylinders are designed for high-pressure, high-flow, high-force applications and corrosive environments.

They are used in a wide range of industrial and mobile applications for pulling and pushing operations, especially when precise low-speed control is required.

The following list highlights some of the application areas:

- Steel mills
- Defense sector
- Automotive plants
- Construction sector
- Chemical and food industries
- Industrial production and assembly jobs
- Mining, offshore, and aerospace applications
- Metal-cutting, machine tools, and forge-pressing machines
- Clamping, pressing, cutting, holding, feeding, lifting, and material handling operations
- Mobile applications for heavy-duty digging, hoisting, tilting, and positioning operations
- Hydraulic cylinders are used in transportation equipment, hoists, and are required for many nautical steering mechanisms in shipbuilding

The growing use of hydraulic cylinders in industrial machines, agricultural machinery, and material-handling equipment for infrastructure development will increase demand for these cylinders. The self-explanatory

Figure 10.1 presents typical examples of hydraulically driven mechanisms.

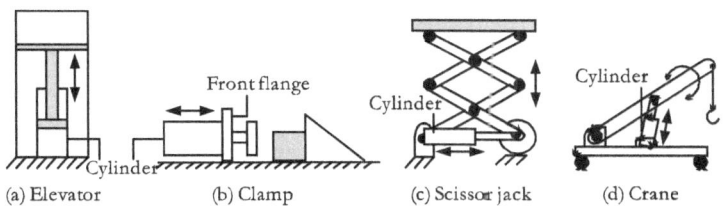

(a) Elevator   (b) Clamp   (c) Scissor jack   (d) Crane

Figure 10.1 | Examples of hydraulically-driven mechanisms

Application of Swing Cylinders

Hydraulic swing clamps clamp workpieces when it is essential to keep the clamping area free of fastening straps and clamping components, enabling unrestricted workpiece loading and unloading. A typical application of swing cylinders is shown in the self-explanatory Figure 10.2.

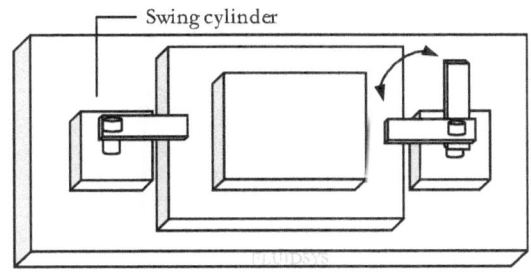

Figure 10.2 | An example of work-piece clamping

Note: Appendix 3 provides key specifications to consider when selecting hydraulic cylinders.

Chapter 11 | Standards, Hydraulic Cylinders

Hydraulic cylinders are available in accordance with ISO and NFPA standards. The ISO hydraulic cylinders are manufactured in accordance with the standards 6020-1, 6020-2, 6020-3, and 6022. The ISO hydraulic cylinders have standard installation dimensions and are therefore interchangeable. Table 11.1 gives some ISO standards for hydraulic cylinders.

Table 11.1 | Important ISO standards for hydraulic cylinders

ISO 6020-1:2007	This standard establishes metric mounting dimensions for medium-series cylinders with bores from 25 mm to 200 mm, for pressures up to 160 bar. The medium series dimensions apply to round head cylinders.
ISO 6020-2:2006	This standard establishes metric mounting dimensions for compact series cylinders with bores from 25 mm to 200 mm for pressures up to 160 bar. The compact series dimensions are most applicable to square head cylinders.
ISO 6020-3:2006	This standard establishes metric mounting dimensions for compact series cylinders with bores from 250 mm to 500 mm for pressures up to 160 bar. The compact series dimensions are most applicable to square head cylinders.
ISO 6022:2006	Establishes mounting dimensions for hydraulic cylinders for use at 250 bar
ISO 3320:1987	This standard establishes a metric series of bores and diameters for fluid power cylinders.
ISO 4395:2009	This standard specifies thread dimensions and configurations for use with hydraulic cylinder piston rod-ends.
ISO 6099:2009	This standard specifies a code for identifying cylinder mounting, envelope, accessory, and connector dimensions, as well as cylinder mounting and accessory types.

Chapter 12 | Maintenance and Safety of Hydraulic Cylinders

When installed and operated in accordance with the manufacturer's instructions, hydraulic cylinders are generally maintenance-free.

The most critical elements of a hydraulic cylinder, from the maintenance point of view, are its seals and piston rods.

The surface of the piston rod that enters the piston rod gland must be kept smooth and clean to avoid the failure of piston rod seals and glands and prevent the wearing of cylinder bearings.

Other issues of concern include air presence, internal and external fluid leaks, and its operation in harsh environmental conditions.

Specialized components, such as metal rod scrapers or protective piston-rod boots, can be installed on the cylinder to protect the piston-rod gland in harsh, dirty environments.

Air contamination, fluid leakage, or the presence of an extremely viscous fluid medium in the cylinder may cause the cylinder to operate sluggishly or erratically. The contaminated fluid inevitably leads to the cylinder's early failure.

Further, the cylinder must be aligned and mounted correctly with its associated load part to reduce side load on the cylinder, avoid consequent failure of cylinder seals, minimise wear of the cylinder bushing, and prevent bending or breakage of the cylinder piston rod.

Impact forces can damage the cylinder's seals.

Ensure that appropriate equipment is available to perform repairs on the cylinder safely if it is heavy.

The following sections explain the various aspects of cylinder maintenance.

Essential Maintenance Activities for Hydraulic Cylinders. Essential hydraulic cylinder maintenance requires routine, proactive care. Apart from the general maintenance activities, the following maintenance activities can be carried out on the cylinder to prevent catastrophic failure, extend service life, and keep it in good working condition:

- Check the piston rod for straightness. Check for any dents or damage on the piston rod due to impact forces
- Examine the piston rod bearing for roundness
- Examine the barrel, the piston, and the piston rod for nicks, scoring, and pitting

- Check the cylinder for worn components
- Check and control the internal and external fluid leakages in the cylinder
- Replace piston seals, piston rod seals, and/or piston rod bushings if leakages occur.

- Align the cylinder and its mating part in line to avoid side loads on the cylinder
- Check the cylinder mountings periodically for tightness or cracks
- Check for sluggish/erratic operation of the cylinder

- Check for the creeping of the cylinder
- Open the bleed ports provided in the cylinder to release the trapped air in the cylinder
- After an initial operating period of typically 40 hours, the screws of the cylinder head and bottom, and all fixing screws, should be retightened to the torque specified
- Ensure lubrication of the bearing points, such as the swivel and articulated bearings, as well as the swivel journal.

Cylinder Faults

Table 12.1 | General faults

Damaged/worn seals	-Replace seals
Excessive rod wear	-Prevent wear
Rod damaged	-Replace the rod
Rod seized	-Repair
Faulty alignment	-Align properly
Broken linkages	-Replace damaged linkages
Loose mountings	-Tighten mounting bolts
Jerky movement	-Assemble piston packing correctly -Avoid heavy load at slow speed
No thrust	-Replace the faulty piston -Prevent excessive leakage -Set PRV to correct pressure
Abnormal cylinder thrust	-Bleed entrained air from the cylinder -Check and set cylinder cushion

Table 12.2 | Erratic operation

Air in the system	-Replace worn seals -Drain and flush fluid -Fill the reservoir to the proper level -Tighten leaking connections -Bleed air -Replace worn shaft/seal
Cylinder/valve sticking/binding	-Remove dirt/gummy deposit -Remove fluid contamination -Remove worn parts -Install seals properly
Cylinder internal leakage	- Repair or replace worn parts -Tighten loose packing -Use medium viscosity fluid -Control fluid contamination -Control wear

Safety Instructions

Many precautions must be taken when installing, handling, or maintaining hydraulic cylinders. Some essential precautions are outlined below. Failure to follow the precautions may result in personal injury, property damage, or both.

This is the safety alert symbol. It is used on machine labels and instruction manuals to alert users to potential personal injury hazards. Therefore, it is essential to follow all safety messages marked with this symbol to avoid injury or death.

- Depressurize a cylinder before repairs are made.
- Any risk of injury should be avoided during the installation and testing of hydraulic cylinders.
- Read the owner's manual completely and familiarize yourself thoroughly with the product and its components, and recognize the hazards associated with its use.
- The owner and operators shall understand the safe operating procedures for cylinders and the system before using them.
- The operator must be trained, qualified, and familiar with the correct operation, maintenance, and use of cylinders and the associated machine.
- Instructions and safety information shall be provided in the operator's native language before the product is used.
- Ensure the operator thoroughly understands the inherent risks associated with the product's use and misuse.
- Wear personal protective equipment (PPE) when installing, operating, or maintaining hydraulic equipment
- Stay clear of a lifted load before it is adequately supported.
- A hydraulic cylinder, when used as a load-lifting device, should never be used as a load-holding device. After the load has been raised, it should be blocked
- Never attempt to lift a load weighing more than the capacity of the cylinder. Overloading can cause equipment failure and personal injury.

- Never rely on hydraulic pressure alone to support a load
- Keep hands and feet away from the cylinder and workpiece during the operation of the cylinder
- Do not subject a cylinder to shock loads
- A cylinder must be able to support the load while pushing or lifting, and hence it must be on a stable base
- Ensure cylinder is fully engaged with adapters and extension accessories
- A cylinder should not be subjected to off-center loads to avoid side loads
- Entrust the repair of hydraulic cylinders to qualified and authorized personnel
- All cylinder fittings should be tightened with proper tools for leak-free connections
- Do not over-tighten the connections. Over-tightening can cause premature thread failure
- Pay attention to the possibility of over-pressurization in cylinder chambers. If required, install additional pressure relief valves or reduce the working pressure
- If larger masses are to be stopped by end-cushion devices, make sure that the cushioning pressure is not too high. The final adjustment must be made on the control side to obtain the most effective cushioning
- The load applied to the piston of a long-stroke cylinder should not exceed the permissible force to avoid the buckling of the piston rod
- Keep hydraulic cylinders away from flames and heat

Summary of Hydraulic Cylinder Maintenance: Cylinder maintenance requires regular, proactive care to prevent downtime, seal failures, and costly damage.

Key actions include keeping hydraulic fluid clean and free of contaminants, checking for external leaks, inspecting rods for corrosion or damage, ensuring proper alignment to prevent uneven wear, and promptly replacing worn seals.

Chapter 13 | Design of Hydraulic Cylinders

Hydraulic cylinder design involves determining the bore diameter, piston rod diameter, stroke length, and material based on load requirements and operating speed. Important factors also include selecting appropriate seals, choosing mounting types, and assessing buckling resistance to maintain structural integrity.

Materials for Cylinder Barrels

The most common material for hydraulic cylinder barrels is low-carbon steel or low-carbon cold-drawn seamless steel tubing, with a tensile yield strength of about 60,000 psi. This material can be used to construct cylinders rated up to 6000 psi and a 5-inch bore. To construct cylinders with higher pressure ratings or larger bore sizes, ductile steel with a higher yield strength should be used. Cast iron should never be used for pressures over 2000 psi, regardless of wall thickness.

Burst Pressure

The wall thickness of a steel tube is considered thin if it is less than 10% of the tube ID, and the wall thickness is considered thick if it is greater than 10% of the tube ID. Barlow's formula can be used to find the burst pressure of a thin-wall steel tube and is given by:

$$\text{Burst pressure (BP)} = 2tS/ID$$

Where,

t = Wall thickness, inch
S = Tensile strength of the tube material, psi
ID = Inside diameter

The following formula can calculate the burst pressure:

$$\text{Burst pressure (BP)} = S \times (R^2 - r^2) / (R^2 + r^2)$$

Where,

S = tensile strength of the tube material, psi
R = Outside radius, inch
r = Inside radius, inch

Working Pressure

The working pressure of a conductor is the safe pressure to which it can be subjected. It is calculated by dividing the conductor's burst pressure by a safety factor.

$$\text{Working pressure (WP)} = \frac{\text{Burst pressure (BP)}}{\text{Safety factor (SF)}}$$

- A safety factor of 4:1 is used for hydraulic applications where shock and mechanical strain are not considerable.
- A safety factor of 6:1 should be used where considerable shock and mechanical strain are expected.
- A safety factor of 8:1 should be used where severe hydraulic shock and mechanical strain are expected.

Example 13.1

A hydraulic cylinder made of mild steel has an inside diameter of 6 inches and a wall thickness of 3/4 inches. What is the maximum working pressure that can be applied to the cylinder if the tensile strength of the material of the cylinder construction is 50000 psi? Assume a safety factor of 5.

Solution

Inside diameter, ID	= 6 inches
Wall thickness, t	= 3/4 = 0.75 inches
Outside diameter, OD	= 6+2x0.75 = 7.5 inches
Tensile strength, S	= 50000 psi
Safety factor	= 5

Inside radius, r	= 6/2 = 3 inches
Outside radius, R	= 7.5/2 = 3.75 inches
Burst pressure	= $Sx(R^2-r^2)/(R^2+r^2)$ [As t >10% of ID]
	=50000 x $(3.75^2 - 3^2)$ / $(3.75^2 + 3^2)$
	=50000 x (14.06 – 9) /(14.06 + 9)
	=50000 x 5.06 /23.06 = 10975 psi
Working pressure	= Burst pressure / Safety factor
	= 10975/5 = 2195 psi

Chapter 14 | Manufacturing Process of Hydraulic Cylinders – Salient Points

Hydraulic cylinders are used in a wide range of machines and systems for industrial and mobile applications. As we know, they operate in high-pressure fluid media. Furthermore, they must withstand high temperatures and stresses in the application environment. This chapter outlines the materials and key processes involved in the manufacturing of hydraulic cylinders.

The manufacturing process of cylinders involves various steps, including heat treatment and machining. The manufacturing process can use in-house machinery such as grinding, boring, drilling, honing, and welding machines. The process may also use additional auxiliary systems for heat treatment, testing, and related processes. Alternatively, many workflows can be outsourced to reduce costs. But remember, outsourcing may degrade cylinder quality.

The manufacture of high-quality cylinders, with good surface quality and geometric accuracy, requires extensive engineering and design capabilities, the selection of high-quality construction materials, high-performance manufacturing systems, and expert workmanship. The following sections present the key points of the hydraulic cylinder manufacturing process for initial understanding.

Steps for Cylinder Manufacturing
The manufacturing process of hydraulic cylinders mainly involves the following steps: design, heat treatment, machining, coating, assembly, and testing.

The Design Phase, Cylinder Manufacturing
The design phase of a cylinder is an important stage in its manufacturing. The design must account for its operating conditions, performance requirements, size constraints, and cost constraints.

In the design stage, the cylinder's requirements are translated into technical drawings that specify materials of construction, dimensions, tolerances, internal-surface quality, and coating methods. The design can leverage various software packages to quickly and accurately prepare drawings.

Parts of a Hydraulic Cylinder

The manufacturing process for a hydraulic cylinder primarily involves preparing and assembling its constituent parts. A hydraulic cylinder typically consists of the following: barrel, piston, piston rod, end plates, seals (piston seals, rod seals, wear bands, wiper seals, etc.), ports, and mounting accessories.

Parameters of Hydraulic Cylinders

The major parameters of a hydraulic cylinder may include the following:
-Inner diameter (Bore diameter)
-Wall thickness
-Piston outer diameter
-Piston rod diameter
-Stroke length
-Depths of seal grooves
-Widths of seal grooves
-Sizes of the seal ring holes
-Roundness
-Plating thickness

Machinery and Equipment for Cylinder Manufacturing

The manufacturing process for hydraulic cylinders requires high precision and specialized machinery to ensure tight tolerances. High-capacity, high-precision CNC lathes, milling machines, boring machines, drilling machines, and welding machines are required to efficiently machine the parts. Highly specialized CNC machines can provide flexibility in machining complex, non-standard hydraulic cylinder components. Other machinery and equipment are also required for operations such as coating, assembly, and testing.

Barrel Preparation

A barrel is the main part of a hydraulic cylinder. The barrel, along with the piston and end covers, should form a sealed chamber for containing the pressurized system fluid. The barrel should be inseparably welded to the cylinder mounting fixtures.

Requirements of Barrel

The barrel should have sufficient strength to prevent deformation under operating and shock pressures. It should also have sufficient rigidity to prevent lateral bending.

The barrel should also have smooth interior surfaces with low roughness, high-precision tolerances, and a durable service life.

Furthermore, the barrel that must be welded to a flange, mounting fixtures, or joints should be weldable to prevent cracking or excessive deformation of the parts after welding.

The requirements for hydraulic cylinder barrels can be met by selecting an appropriate barrel material. Hydraulic cylinder barrels are primarily constructed from high-strength, cold-drawn or hot-rolled seamless steel tubing to withstand high internal pressures and wear.

Common materials include carbon steels for standard applications, alloy steels for high-pressure needs, and stainless steels for corrosive environments. The barrel materials should have high strength, high wear and corrosion resistance, good machinability, good weldability, and low cost.

Heat Treatment

Heat treatment is an essential process in the manufacturing of hydraulic cylinders to enhance their strength, durability, and dimensional stability. It involves heating a metal part to a specified temperature, holding (soaking) it at that temperature for a specified duration (soak time), and then cooling it at a controlled rate using methods to achieve the desired properties.

During heat treatment, the metal part changes its physical structure and mechanical properties, including hardness, ductility, brittleness, and corrosion resistance. Some of the most common heat treatment methods include annealing, hardening, tempering, carburization, etc.

In the **annealing process**, the metal is heated above its upper critical temperature and then cooled slowly. Annealing is performed to soften the metal, improving its machinability, ductility, and toughness.

In the **hardening process**, a metal part is heated to a specified temperature and then rapidly cooled (quenched) by submerging it in a cooling medium such as oil, brine, or water. This process is performed to increase the part's hardness. However, the part can become more brittle.

In the **tempering process**, temperatures are typically set well below the hardening temperature. This process is carried out to reduce excess hardness and, therefore, brittleness induced during hardening.

In the **carburization process**, the metal part is heated in the presence of a carbon-containing gas, such as carbon monoxide, which releases carbon upon decomposition. The released carbon is absorbed into the metal surface. The surface of the metal part gets hardened as the carbon content of the surface increases.

Types of Stainless Steel
Stainless steel is an alloy of steel that contains chromium, nickel, molybdenum, and/or titanium. Adding elements to steel improves its properties, such as hardness, ductility, and corrosion resistance.

For example, the primary role of Cr in steel is to improve properties such as strength, hardness, and impact toughness after quenching and tempering.

Stainless steel can be classified into one of five types. They are austenitic, ferritic, martensitic, duplex, and precipitation. Each of these types can again be subdivided into grades of stainless steel.

The stainless steel families are specified by the ratio of the metals that compose the alloy. An example of a commonly used grading system is the American Iron and Steel Institute (AISI) system. Each category is further divided into series and grades. The grades reflect the specific alloy's durability, quality, and temperature resistance.

The high chromium content in austenitic steels makes them a good choice for corrosion resistance. Stainless steel alloys containing chromium and nickel often provide the best combination of strength, ductility, and toughness.

Materials for Barrel
There are many steel grades available, classified by tensile strength. A hydraulic cylinder barrel can generally be prepared from an annealed, cold-drawn, or hot-rolled seamless steel tube, or from a forged piece. The steel type is usually carbon steel. However, in applications with corrosive environments, stainless steel can be used. The barrel with a thick wall should be made with cast iron or forgings.

The commonly used materials include:
-Carbon steel: #20 steel, #35 steel, and #45 steel, St 52
-Low tensile carbon steel: AISI 1020
-Medium tensile carbon steel: AISI 1045
-Ordinary structural low-alloy steel: 15MnV and 27SiMn
-Structural alloy steel: 30CrMo, 35CrMo, and 35CrMo-ALA
-Stainless steel: Cr18Ni9
-Low alloy steel: AISI 4140
-Austenitic steel: AISI 304 (18 Cr/8 Ni), AISI 316
-Ferritic steel: AISI 409, 430, 439, and 441
-Duplex grade steel: 2205 (22%Cr + 5%Ni)
-Cast steel: ZG230-450 and ZG310-500

Inspection of Materials for Barrel

It is essential to inspect the tube materials for any defects. Test certificates issued by manufacturers for the chemical composition and mechanical properties of their products must be verified as suitable.

Rough Turning and Other Operations, Barrel Preparation

First, cut the tube to the correct length during parting. Then, turn the barrel's outer surface to the required diameter and shape. Furthermore, chamfer the barrel to remove sharp edges and add the required external threads.

Machining Process (Boring) for Barrel

Machining is a metal removal process that depends on boring, cutting, and grinding operations to remove unwanted material from the barrel to achieve a final shape. Before boring, place the barrel in the boring machine holder and secure it. Use bolts to tighten and adjust the height of the boring tooltip until it aligns with the barrel center.

Perform the boring operation to finish the inner surface of the semi-finished barrel. The boring operation is the main process of machining the barrel. It is essentially a turning operation performed on the internal surfaces of the barrel to achieve dimensional and surface finish tolerances, thereby realizing high dimensional accuracy.

Precision Machining, Barrel

After cold drawing and heat treatment, the partly finished barrel is additionally prepared for improving the surface finish and the geometric form of the inner surface of the barrel using the following methods:

(1) Honing process and

(2) SRB (Skiving & Roller Burnishing) process.

Honing Process

During honing, the inside surface of a barrel is ground with abrasive stones and paper. The polishing stone turns while being moved in and out of the barrel. This process can remove small amounts of material from the inner surface, remove surface imperfections, and achieve precise inside-diameter dimensions and tolerances, providing optimal lubrication and low wear on cylinders and seals.

It may be noted that ID dimensional tolerances as precise as 0.021mm (H7) and surface roughness in the range Ra 0.2 to 0.8 um can be achieved for the barrel using the honing process.

Skiving and Roller Burnishing (SRB) Process

The skiving and roller burnishing (SRB) process uses a skiving tool on the forward stroke to skive the barrel's inside diameter to the required size. It uses roller bearings on the return stroke to burnish the surface to the required roughness. Compared with the honing process, the SRB process is faster and can achieve more precise tolerances and a smoother surface finish.

Honed steel tube manufactured with the SRB process typically has an ID dimension tolerance of 0.021mm (H7) and an inside-surface roughness of Ra < 0.4 μm.

Precision Turning, Barrel Preparation

Turn the outer barrel circle to the correct size and shape. Drill the fluid ports if the ports are designed to be on the barrel.

Welding, Barrel Preparation

Weld pipe joint seats, flange, and other accessories.

Final stages of Barrel Preparation

During the final stages of barrel preparation, it is brushed with oil to prevent rust. Then, it is moved to a location with all precautions for safe storage.

Preparing the Piston

The piston is first machined to the correct size. It must then be quenched and tempered for hardness. Verify its hardness through a hardness test.

The correctly-sized piston is, then, machined with grooves to fit seals and bearing elements. The piston is inseparably attached to the piston rod using threads, bolts, or nuts. The piston must be coated with non-ferrous metal to ensure corrosion resistance and precise guidance.

Preparing the Piston rod

The piston rod is typically made of steel or stainless steel. Check the material certificate furnished by the manufacturer for its suitability for the piston rod.

The piston rod must be machined to obtain the correct shape and size. First, it must be roughly machined. It must then be quenched and tempered for hardness. Verify its hardness through a hardness test.

It must then be heat-treated through induction hardening, carburizing, or nitriding to improve its surface hardness.

Induction hardening can improve the piston rod's mechanical properties, resist buckling during push operations, and protect it from damage in the event of external impact.

It must then be precision-machined, finely ground, and polished to provide a reliable seal and prevent leakage. The rod ends must be finished with rounded edges.

The piston rod is then either hard chrome-plated or subjected to surface heat treatment. It must then be polished to provide a reliable seal and prevent leakage. The piston rod needs to be highly resistant to pitting, corrosion, and wear.

Check the prepared piston rod for hardness and dimensional accuracy. Coat the piston rod with oil to prevent rust during storage.

Preparing the Piston Guide Sleeve

Check the certificate for the piston guide sleeve material to verify suitability. Perform rough machining, then precise machining, to achieve the correct sleeve size. Inspect the prepared sleeve. Coat the sleeve with oil to prevent rust and for storage.

Preparing the Cylinder Head

The manufacturing process for a cylinder head with rod sealing, guiding, cushion, and porting arrangements includes inspection of the material certificate, rough machining, quenching and tempering, hardness testing, precision machining, internal and external threading, sawing, milling, boring and fitting, and final inspection.

The head is connected to the body using threading, bolts, or tie rods. An O-ring seal is used between the head and barrel.

Cylinder Ports

A port is an important attachment in a cylinder. It permits fluid flow both into and out of the cylinder. Furthermore, it must be secure, as a vulnerable port can cause a dangerous hydraulic fluid leak under high pressure.

Usually, cylinders have standard port sizes and locations. The standard port locations are on the top of the head and cap. However, it is possible to select different port locations to accommodate plumbing or other location-specific requirements. Many porting options support a wide range of cylinder customizations.

On Hydraulic Cylinders, standard ports are SAE O-ring threaded ports, also referred to simply as SAE ports. SAE ports are manufactured in accordance with SAE J1926-1. Another porting option is a four-bolt flange port. These ports are manufactured to SAE J518, code 61 and 62 specifications. Other port options include NPTF dry-seal threads and BSPP threads.

Coating, Painting, and Polishing, Barrel
The exterior surface of the barrel must be spray-painted in accordance with the relevant standards.

The internal surface of the barrel usually requires no paint or chrome because the hydraulic fluid protects it from corrosion. If the hydraulic fluid is corrosive (e.g., water), the inside of the tube can be coated.

Hard chrome plating can increase hardness, reduce friction, and improve wear resistance. Nickel phosphorus coating is suitable for corrosive, high-wear environments, such as marine applications. Nitridation or carburization can improve surface hardness and wear resistance.

Coating, Piston
Hard chrome plating is applied to the piston.

Coating, Piston rods
The surfaces of piston rods are often treated with techniques such as Nickel-Chromium plating, laser cladding, supersonic flame spraying, or thermal spraying to enhance wear and corrosion resistance.

Coating, Further Explanation
Chromium plating is the most common surface treatment for piston rods. However, chrome plating can develop cracks and cause corrosion. Corrosion resistance can be improved by applying a nickel underlayer. The coatings can then be finished to the desired surface roughness for optimum sealing performance. A coating technique can be selected based on specific operational wear conditions, such as high impact or abrasive forces, corrosion, and hot environments.

Seals

Seals are used in a cylinder to contain the system fluid and prevent leakage. They fall into two main types. That is dynamic and static seals.

Static seals are used when there is no relative motion between mating surfaces.

Dynamic seals are used between moving parts. Dynamic piston seals are used between the piston and the cylinder bore to accommodate reciprocating motion. Dynamic piston rod seals are used between the piston rod and head to mitigate dynamic stresses.

Selecting the right seal profile and material for a given application requires consideration of factors such as piston rod and bore diameters, seal groove dimensions, and gaps.

Assembly

Hydraulic cylinder assembly involves precisely installing seals, bearings, and the piston onto the rod, then carefully inserting the rod assembly into the barrel. Key steps include lubricating seals, cleaning components, securing the cylinder, and tightening end caps or gland nuts to the specified torque.

Preparation: The barrel, piston rod, piston, and gland must be thoroughly cleaned to prevent contamination.

Seal Installation: Piston and piston rod seals are lubricated and installed. Care is taken to prevent damage during seal installation.

Rod Assembly: The piston is secured to the piston rod.

Insertion: The rod assembly is carefully inserted into the cylinder barrel.

Closing the Cylinder: The gland (head end) is installed and tightened to the tube. The cap is secured, or, in the case of tie-rod cylinders, the rods are tightened in a cross pattern to ensure even pressure.

Final Inspection: The cylinder is inspected and tested for performance.

Quality Control

A strict quality-control procedure should be in place at every step of hydraulic cylinder production to ensure compliance with specifications. The quality of the hydraulic cylinder is assured by the superior manufacturing processes, the quality of raw materials, including metals and seals, used for the cylinder, the type of checks that are performed, and the expert quality.

Tests and Inspections

The manufactured cylinder must undergo performance testing under load to confirm compliance with the specified requirements. Ensure that all cylinder dimensions and technical requirements comply with the drawing. Arrange inspection by third-party agencies if required. The following tests or inspections may be carried out on the cylinder:

-Visual inspection
-Outer diameter inspection
-Bore diameter inspection
-Stroke Length
-Surface roughness inspection
-Finish test

-Pressure test
-Load test
-Hardness test
-Paint layer test
-Plating thickness (>0.04 mm)
-Concentricity inspection

-Ultrasonic test
-Penetrant test of the welds
-Destructive test for finding weld strength
-Oil port penetration inspection
-Endurance test
-Internal leakage test
-Operational Smoothness

Brief Explanation of Some Tests

A brief explanation of some of the tests is given below:

Visual inspection: Check for dust, scratches, dents, uneven paint, color differences, bending, sagging, and smooth plating. Examine the cylinder body and welds for cracks or damage.

Hardness Test: The material's hardness can be measured using a hardness test.

Endurance Test: Under rated pressure, the hydraulic cylinder is operated continuously at the design-required maximum speed for more than 8 hours. Operating the hydraulic cylinder through multiple cycles, typically up to 50,000, to ensure consistent performance, often followed by re-inspection.

Internal Leakage Test: Perform the test as follows. Direct hydraulic fluid at the nominal pressure to the working chamber of the hydraulic cylinder, and measure the leakage from the working chamber to the unpressurized chamber.

For example, a common method involves extending the cylinder, disconnecting the hose at the piston-rod-side port, and pressurizing the piston-side port. Oil exiting the open port indicates piston seal bypass.

External Leakage Test: Measure leakage at the piston rod seal and the joint.

Pressure Test: A cylinder is tested at 1.5-2.5 times its working pressure to assess the structural integrity of the cylinder body under high pressure.

Operational Smoothness: Inspect for any abnormal sounds or irregular movements during operation.

15 | Objective Type Questions

1. What determines the speed of a cylinder used in a hydraulic system?
   a) Fluid flow rate
   b) System pressure
   c) Size of the applied load
   d) Stroke length

2. Wear bands in hydraulic cylinders eliminate:
   a) impact forces
   b) cylinder scoring
   c) leakage
   d) pressure drop

3. Which type of hydraulic cylinders is used to multiply force in a limited lateral space?
   a) Duplex cylinder
   b) Telescopic cylinder
   c) Tandem cylinder
   d) Ram cylinder

4. Which type of hydraulic cylinders is most suitable for the harsh service conditions?
   a) Tie-rod cylinders
   b) Mill cylinders
   c) Threaded-end cylinders
   d) Welded cylinders

5. Which is a suitable material for hydraulic cylinder barrel?
   a) Aluminium
   b) Anodized aluminium
   c) Carbon steel
   d) Polyurethene

16 | Review Questions

1. Define the term hydraulic actuator

2. Explain the function of a hydraulic linear actuator by taking a double-acting cylinder as an example.

3. Name two basic types of hydraulic actuators and differentiate them.

4. What are the different types of loads associated with hydraulic systems? Explain briefly.

5. Describe the relationship between the fluid flow rate, piston area, and piston velocity of a hydraulic cylinder.

6. Why does a hydraulic double-acting cylinder retract at a higher speed than it does while extending for the same input flow rate?

7. Describe the construction and design features of standard hydraulic cylinders briefly.

8. Explain the operational and other features of different types of hydraulic cylinders.

9. Name the most general types of hydraulic cylinders.

10. What are the problems encountered by the tie-rod cylinders because of their body style?

11. What are the key advantages of welded-style hydraulic cylinders?

12. Describe the constructional features of hydraulic cylinders

13. Explain the purpose of the piston rod boot in a hydraulic cylinder.

14. Explain the purpose of the stop tube in a long-stroke hydraulic cylinder.

15. What are the different types of seal materials used in hydraulic actuators?

16. Name four different types of mounting arrangements for hydraulic cylinders.

17. Give a brief account of the commonly used hydraulic actuators.

18. Describe hydraulic actuators with neat sketches.

19. What are the different methods to retract single-acting cylinders?

20. What is the difference between the single-acting and the double-acting hydraulic cylinders?

21. What is a double-rod cylinder? When is it used?

22. Why does the piston rod of a double-acting cylinder retract at a higher velocity than when it extends for the same input flow rate?

23. What is the differential cylinder effect of a conventional hydraulic cylinder?

24. What is cylinder cushioning? Explain with a diagram.

25. Why are end cushions used in hydraulic cylinders?

26. Explain the working of a telescopic hydraulic cylinder, with a neat diagram.

27. What is a telescopic hydraulic cylinder? When is it employed?

28. Briefly explain the working of the hydraulic actuators: (1) Tandem cylinder and (2) Duplex cylinder

29. List some advantages of hydraulic cylinders.

17 | Numerical Problems

1. A 5.12 in bore hydraulic cylinder is to lift a load of 161862 lb. What operating pressure is required to lift the load? {Ans: 7861 psi}

2. A 1.77 in bore hydraulic cylinder on a construction site is used to lift a load. The gauge shows an operating pressure of 4931 psi. What is the weight of the load? {Ans:12130 lb}

3. What is the speed of a hydraulic cylinder with a piston area of 20.46 in^2 supplied with fluid at the rate of 101.7 in3/s? {Ans: 5 in/s}

4. A single-acting hydraulic cylinder has an active area of 0.02185 ft^2 and a stroke of 0.328 ft. How much fluid does the cylinder require per cycle? {Ans: 0.007165 ft^3}

5. At what speed will a hydraulic cylinder, with an effective area of 4.65 in^2, move when powered by a pump with a no-load flow rate of 6.1 in^3/s? {Ans: 1.31 in/s}

6. Calculate the forward speed of a double-acting hydraulic cylinder with a bore diameter of 3.281 ft and a flow rate of 6102 in^3/min. {Ans: 0.084 in/s}

7. A double-acting hydraulic cylinder with a single piston rod, used for lifting and lowering heavy loads in a marine application, must produce a thrust of 17985 lb and move out with a velocity of 0.0098 ft/s on the out-stroke. The operating pressure is 1450 psi. Calculate the cylinder bore diameter and the required flow rate.

{Ans: 3.97 in, 1.46 in^3/s}

8. A single-acting hydraulic cylinder has a piston of 2.95 in diameter and is supplied with fluid at 1160 psi and a flow rate of 0.562 cfm. Calculate the thrust, velocity, and power.

{Ans: 7923 lb, 2.37 in/s, 2.845 hp}

9. A double-acting hydraulic cylinder is used to reciprocate in an application. The relief valve setting is 1015 psi. The piston area is 24.955 in^2, and the rod area is 6.99 in^2. If the pump flow is 85.43 in^3/s, find the speed and load-carrying capacity of the cylinder for its: (i) extension stroke and (ii) retraction stroke.

{Ans: 3.42 in/s, 12.22 in/s, 25329 lb, 7095 lb}

10. A hydraulic cylinder has a bore diameter of 3.1496 in and a piston rod diameter of 1.5748 in. If the cylinder receives a flow of 101.7 in^3/s at 1740 psi pressure, find: a) extension as well as retraction speeds, and b) extension as well as retraction load-carrying capacities of the cylinder.

{Ans: 13.06 in/s, 26.07 in/s, 13555 lb, 6786 lb}

11. A 3 in diameter hydraulic cylinder has a 1.5 in diameter piston rod. What is the flow rate leaving through the piston rod port of the extending cylinder when the flow rate that enters its piston port is 8 gpm?

{Ans: 32 gpm}

12. A 3¼ inch diameter hydraulic double-acting cylinder has a 1¾ diameter rod. If the cylinder receives flow at 30 gpm/115.5 in^3/s and 1740 psi, find the (1) thrust and pull forces produced by the cylinder, and (2) extension and retraction speeds of the cylinder, and compare.

{Ans: 14442 lb, 1.16 ft/s, 1.63 ft/s}

Objective-type questions - answer key:
1.   *1-a, 2-b, 3-c, 4-b, 5-c*

Appendix 1

Bore diameters and Piston rod diameters in the English units

Bore (inch)	Rod Diameter (inch)
1.0	0.500
	0.625
1.5	0.625
	1.000
2.0	0.625
	1.375
	1.000
2.5	1.000
	1.750
	1.375
	0.625
3.25	1.000
	2.000
	1.375
	1.750
4.0	1.375
	2.500
	1.750
	2.000
	1.000
5.0	1.750
	3.500
	2.000
	2.500
	3.000
	1.000
6.0	1.750
	4.000
8.0	2.000
	5.500

Appendix 2

Theoretical Cylinder Forces in the English Units
Force (lb) = Pressure (psi) x Piston Area (in²)

Table A2.1: Theoretical Cylinder Forces in the English Units

Bore size (in)	Rod dia. (in)	Force (lb)	System pressure (psi)			
			1000	2000	3000	5000
1½	¾	Thrust	1,770	3,530	5,300	8,830
		Pull	1,320	2,650	3,970	6,630
2	1	Thrust	3,140	6,280	9,420	15,700
		Pull	2,360	4,710	7,070	11,780
3	2	Thrust	7,070	14,130	21,200	35,340
		Pull	3,930	7,850	11,780	19,630
4	3	Thrust	12,550	25,130	37,700	62,830
		Pull	5,500	11,000	16,490	27,490
6	4	Thrust	28,270	56,550	84,820	141,370
		Pull	15,700	31,410	47,120	78,540
8	5	Thrust	50,260	100,530	150,800	251,330
		Pull	30,630	61,260	91,890	153,150
10	6	Thrust	78,540	158,080	235,620	392,700
		Pull	50,260	100,530	150,800	251,330

Example

Bore diameter	$= 6$ in
Rod diameter	$= 4$ in
Pressure, P	$= 1000$ psi
Piston area, A_{ext}	$= \prod D^2/4 = 3.1416 \times 6^2/4 = 28.27$ in²
Effective area for pull stroke, $A_{ret} = \prod (D^2 - d^2)/4$	
	$= 3.14 (6^2 - 4^2)/4 = 15.7$ in²
Thrust	$= $ P x Aext
	$= 1000 \times 28.26$
	$= 28270$ lb
Pull	$= $ P x Aret $= 1000 \times 15.7$
	$= 15700$ lb

Appendix 3

Different types of Piston Seals
- Lip Seal
- Spring-Loaded PTFE Seal
- Magnetic Piston, stainless steel cylinder body, single bi-directional piston seal
- Magnetic Piston, carbon steel body, single bi-directional piston seal
- Magnetic Piston, Aluminum Tube

Ports
- SAE Straight Thread O-Ring
- NPTF Ports (Dry Seal Pipe Thread)
BSP Ports (Parallel Thread ISO 228)
BSPT Ports (Taper Thread)
Metric Thread Ports
Metric Thread Ports per ISO 6149

Mounting Styles
- Tie Rods Extended Head-end
- Tie Rods Extended Cap-end
- Tie Rods Extended at Both Ends
- Head Rectangular Flange
- Head Square Flange
- Cap Rectangular Flange
- Cap Square Flange
- Side Lug
- Side Tapped
- Cap Fixed Clevis
- Head Trunnion
- Cap Trunnion
- Intermediate Fixed Trunnion
- Spherical Bearing

Appendix 4

Important specifications to be considered while selecting hydraulic cylinders

Table A4.1 | Important specifications of hydraulic cylinders

Bore size	Should satisfy the requirement of thrust/pull for the operating pressure
Piston rod diameter	Should prevent piston rod buckling
Single rod / Double rod	
Cushions	Yes / No If yes, Head-end, Cap-end, or both ends?
Stop tube	Yes / No
Piston and piston rod Seal type	Fluid and temperature compatibility?
Stroke length	
Piston rod-end thread style	
Port size	For a given speed requirement
Port position	
Mounting style	
Piston rod and mounting accessories	To attach the cylinder to the load
Optional accessories	
Fluid medium	

Appendix 5

Seal Materials and their Temperature Ranges

Table A5.1

Material	Temperature Range
Nitrile	-22°F to 212°F
H-Nitrile (Hydrogenated Nitrile)	-30°F to 302°F
Viton	-4°F to 400 °F
Silicone	-75°F to 450°F
EPDM	-65°F to 350°F
Polyurethane	-22°F to 230°F
Nylon	-40°F to 248°F
Teflon, virgin	-328°F to 500°F
Teflon, filled	-328°F to 500°F

Appendix 6

Basics of Threads

Hydraulic pipe threads, as shown in Figure 2.6, are specialized threaded connections that create leak-free, high-pressure seals in hydraulic systems. A screw has a male thread (external), while the matching hole has a female thread (internal). Pipe threads are classified as either tapered or parallel (straight).

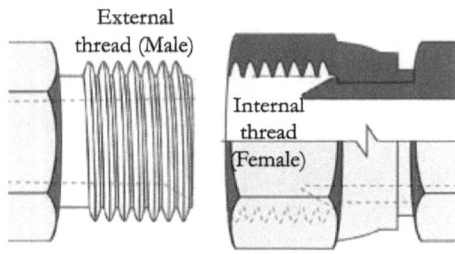

Figure 2.6 | Male and female threads

Parallel Threads and Tapered Threads

Figure 2.7 shows parallel and tapered threads. Parallel threads maintain a constant diameter along their length. Sealing in parallel threads is achieved via O-rings.

Tapered threads gradually decrease in diameter as they extend outward. Sealing in tapered threads is achieved by deforming the threads.

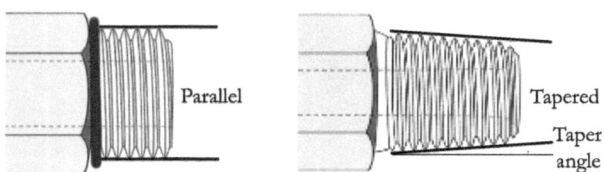

Figure 2.7 | Parallel thread and tapered thread

Terms and Definitions, Threads

Threads are governed by terms such as crest, root, threads per inch (TPI), major diameter, minor diameter, pitch diameter, flank angle, and taper angle. These terms are briefly explained below and illustrated in Figure 2.8.

Crest: The outermost part of a thread is called the crest.

Root: The innermost part of a thread is called the root.

Pitch: The pitch is the distance from the crest of one thread to the next, measured in mm.

Threads per Inch (TPI): Threads per inch is the number of thread peaks per inch of the screw.

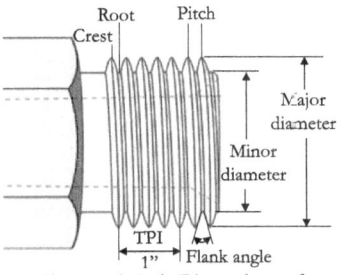

Figure 2.8 | Pipe thread

Major Diameter: The major diameter is defined by the thread tips.

Minor Diameter: The minor diameter is determined by the thread groove

Flank Angle: The flank angle is the angle between the flank of a screw thread and the perpendicular to the screw's axis.

Taper Angle: Tapered threads have a taper angle. This is the angle between the taper and the pipe's center axis.

77

Hydraulic Pipe Thread Types

Hydraulic pipe thread types include ISO coarse and fine threads (metric), British Standard Pipe (BSP) parallel and tapered, and National Pipe Thread (NPT). A type is chosen based on region and pressure requirements. These types and ISO metric screw threads are described below.

M - ISO Coarse Screw Threads (Metric)

The ISO metric screw thread is a globally standardized thread for fasteners and ports, as specified in ISO 724. An ISO metric thread can be coarse or fine. Table 2.3 presents the size chart for ISO metric coarse threads (M-series).

Table 2.3 | Size chart for M- ISO coarse screw thread

Thread size (mm)	Major diameter (mm)	Minor Diameter (mm)	Pitch (mm)
M3	2.98	2.459	0.5
M4	3.978	3.242	0.7
M5	4.976	4.134	0.8
M6	5.974	4.917	1.0
M8	7.974	6.917	1.0
M10	9.968	8.376	1.5
M12	11.97	10.106	1.75
M16	15.96	13.835	2.0
M20	19.96	17.294	2.5
M24	23.95	20.752	3.0

M - ISO Fine Screw Threads (Metric)

An ISO fine-pipe thread is a standardised metric thread with a smaller pitch than a coarse thread, used for precise connections and often designated as M-series. Table 2.4 presents the size chart for ISO Metric fine threads.

Table 2.4 | Size chart for M- ISO fine screw thread

Thread size (mm)	Major diameter (mm)	Minor Diameter (mm)	Pitch (mm)
M3x0.35	2.981	2.521	0.35
M4x0.5	3.978	3.242	0.5
M5x0.5	4.98	4.459	0.5
M6x0.75	5.978	5.188	0.75
M8x0.75	7.978	7 188	0.75
M10x0.75	9.978	9.188	0.75
M10x1	9.974	8.917	1.0
M10x1.25	9.972	8.647	1.25
M12x1	11.97	10.917	1.0
M12x1.5	11.97	10.376	1.5
M16x1	15.97	14.917	1.0
M16x1.5	15.97	14.376	1.5
M20x1	19.97	18.917	1.0
M20x1.5	19.97	18.376	1.5
M20x2	19.97	17.835	2.0
M24x1.5	23.97	22.376	1.5

ISO Metric (M-Series) Threads Notation

The metric designation of a screw thread is indicated by the letter M, followed by the nominal diameter in mm and then the pitch in mm—for example, M 20 x 1.5. For coarse thread, the pitch may or may not be displayed, whereas for fine thread, the pitch must be displayed.

BSPP Threads

The common types of British Standard Pipe (Whitworth) threads are BSPP(G)-Parallel and BSPT (R/Rp)-internally tapered/parallel. An appropriate sealing compound can be applied to the thread to ensure a leak-proof joint. Table 2.5 presents the BSPP (G) size chart (ISO 228).

Table 2.5 | Size chart for BSPP (G) threads

Thread Size (inch)	Major diameter (mm)	Minor Diameter (mm)	Threads Per Inch (TPI)
G 1/16	7.723	6.561	28
G 1/8	9.728	8.566	28
G 1/4	13.157	11.445	19
G 3/8	16.662	14.950	19
G 1/2	20.955	18.631	14
G 3/4	26.441	24.117	14
G 1	33.249	30.291	11
G 2	59.614	56.656	11

BSPT Threads

Table 2.6 presents the BSPT thread-size chart (ISO 7).

Table 2.6 | Size chart for BSPT threads

Male Thread Size (inch)	Female Thread Size (inch)	Major diameter (mm)	Minor Female Diameter (mm)	Threads Per Inch (TPI)
R 1/16	Rp 1/16	7.723	6.490	28
R 1/8	Rp 1/8	9.728	8.495	28
R 1/4	Rp 1/4	13.157	11.341	19
R 3/8	Rp 3/8	16.662	14.846	19
R 1/2	Rp 1/2	20.955	18.489	14
R 3/4	Rp 3/4	26.441	23.975	14
R 1	Rp 1	33.249	30.111	11

NPT Pipe Threads

National Pipe thread is the USA standard for tapered threads. The most common types of National Pipe Threads are National Pipe Taper (NPT) and National Pipe Taper Fuel (NPTF). NPTF is the taper pipe thread for a dry-seal joint without sealant compound. The NPT pipe thread size chart is given in Table 2.7.

Table 2.7 | NPT threads size chart

Thread Size	Major Diameter (mm)	Threads Per Inch (TPI)
1/16" – 27 NPT	7.938	27
1/8" – 27 NPT	10.287	27
1/4" – 18 NPT	13.716	18
3/8" – 18 NPT	17.145	18
1/2" – 14 NPT	21.336	14
3/4" – 14 NPT	26.670	14

Flank Angle

The flank angles for ISO, BSPP, BSPT, and NPT are shown in Figure 2.9.

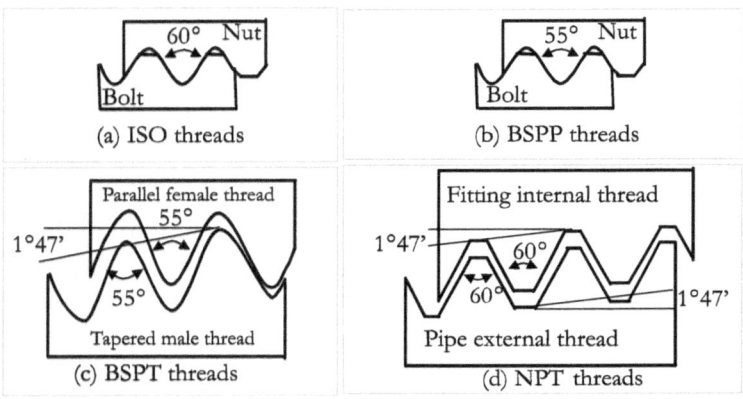

Figure 2.9 | Flank angles

Standards

Important standards for various thread types are given in Table 2.8.

Table 2.8 | Standards pertinent to threads.

Threads	Designation	Type	Standard
M - ISO screw thread (Metric)	M	Coarse thread	ISO 724 (DIN 13-1)
	M	Fine thread	ISO 724 (DIN 13-2 to 11)
NPT – Pipe thread	NPT		ANSI B1.20.1
	NPTF		ANSI B1.20.3
G/R/R_P – Whitworth pipe thread	G	BSPP	ISO 228 (DIN 259)
	R / Rp	BSPT	ISO 7 (EN10226)
UNC/UNF – Unified National Thread	UNC		ANSI B1.1

18 | References

1. Andrew Parr, Hydraulics & Pneumatics, A technician's and engineer's guide, 2nd Edition, Butterworth, Heinemann, 1998

2. Anthony Esposito, Fluid Power with Applications, 6th Edition, Prentice-Hall of India, 2006

3. Article on '8 Steps You Should Follow To Manufacture A High-quality Hydraulic Cylinder', by Antony, Taizhou Chuanhu Hydraulic Machinery Co., Ltd, Zhejiang Province, China

4. Article on 'A Brief Look At The Hydraulic Cylinder Manufacturing Process', Ranger Caradoc Hydraulics Ltd., West Bromwich, United Kingdom

5. Article on 'Clamping elements', HYDROKOMP®, Mücke, Germany, www.hydrokomp.de

6. Article on 'Custom Hydraulic Cylinder Manufacturing', CATSU Hydraulic Machinery Equipment Co., Ltd.

7. Article on 'Cylinders-Part 1', Hydraulics & Pneumatics Magazine, Penton Media, Inc

8. Article on 'FLUID POWER Design Data Sheet' published by WOMAC EDUCATIONAL PUBLICATIONS, Womac Machine Supply Co., 13835 Senlac Dr Farmers Branch, TX 75234

9. Article on 'How to Determine what kind of Porting you will need for your Hydraulic Cylinder Application'' SHEFFER PNEUMATIC AND HYDRAULIC CYLINDERS

10. Article on 'HYDRAULIC CLAMPING', VEKTEK LLC, U.S.A., www.vektek.com

11. Article on 'Hydraulic cylinder anatomy', by Frank J. Bartos, Control Engineering Resource Center

12. Article on 'Hydraulic cylinder', HYDRAULIC EQUIPMENT & TOOLS MARKETPLACE, Hydraulic Equipment Manufacturers

13. Article on 'Hydraulic Cylinders', INTEGRAL HYDRAULIK GmbH & Co. KG Hanns-Martin-Schleyer-Str. 20, 47877 Willich Postfach 500 209

14. Article on 'Hydraulic swing clamp', KOSMEK, Japan

15. Article on 'Hydraulic Swing Cylinders – Top Mounted, Double Acting', Precision Engineering Accessories, Bangalore, India, www.preacindia.com

16. Article on 'Industrial hydraulic cylinders', Herbert Hänchen GmbH

17. Article on 'Mini Swing Clamps with Sturdy Swing Mechanism, threaded-body type, double acting, max. operating pressure 150 bar', Issue B1.848/12-10E, ROEMHELD GmbH

18. Article on 'Requirements and material section of hydraulic cylinder barrel', Wuxi Longzhichen Machinery Co., Ltd., https://www.honedtube.com/info/requirements-and-material-section-of-hydraulic-18363999.html

19. Articles on 'Hydraulic Cylinder Seal Selection', 'Hydraulic Cylinder Repair Tutorial', 'The Advantages of Hydraulic Cylinders', 'Hydraulic Cylinders- Design Considerations for Hydraulic Cylinders', 'Hydraulic Cylinders-Engineering and Design Tips', 'Hydraulic Oil Tutorial, Hydraulic Cylinder Safety Tutorial', and 'Advantages of Welded Body Hydraulic Cylinders', HYCO ULTRAMETAL, Kitchener, Strasburg Road, Unit C Kitchener, Ontario

20. Brochure on 'Cylinders Capabilities', Document No. M-CYOV-MR001-E, January 2007, Eaton, USA

21. Catalogue on 'High-pressure hydraulics', EURO PRESS PACK, Via M. Disma, Carasco (GE), ITALY

22. Catalogue on 'Hydraulic cylinder, Type CDL1, Series 1X, Nominal pressure 160 bar (16 MPa), RE 17325/09.05', Bosch Rexroth AG Hydraulics

23. Catalogue on 'Medium Duty Hydraulic Cylinders, Series 3L', 4/10 / No. HY08-1130-1/NA, Parker Hannifin Corporation, Cylinder Division, USA. www.parker.com/cylinder

24. Catalogue on 'Metric Hydraulic Cylinders, Series HMI, Conforms to ISO 6020/2 (1991), for working pressures up to 210 bar', Parker Hannifin Corporation, Cylinder Division, USA

25. Catalogue on Series H, Milwaukee Cylinder, Milwaukee, USA, www.milwaukeecylinder.com

26. Document on 'OVERVIEW AND SELECTION GUIDE' Doc. no. 941186/Mat. No. 267436 EN, Balluff Inc. 8125 Holton Drive Florence, KY 41042

27. Hydraulic Cylinder Troubleshooting - Parker Hannifin. (n.d.). Retrieved from http://parker.com/literature/Literature%20Files/euro_cylinder/v4/Trouble_1242-1-gb.pdf_br

28. Vocational Training Course, HYDRAULICS – 21 Exercises with Instructions, published by Bundesinstitut fur Berufsbildungsforschung, Berlin, 1973

29. William D. Wolansky et al., Fundamentals of Fluid Power, Houghton Mifflin Company, Boston, 1977

Fluid Power Educational Series Books

Pneumatic Systems and Circuits -Basic Level (In the SI Units)
2. Industrial Pneumatics -Basic Level (In the English Units)
3. Pneumatic Systems and Circuits -Advanced Level
4. Electro-Pneumatics and Automation
5. Design of Pneumatic Systems (In the SI Units)
6. Design Concepts in Pneumatic Systems (In the English Units)
7. Maintenance, Troubleshooting, and Safety in Pneumatic Systems
8. Industrial Hydraulic Systems and Circuits -Basic Level (In the SI Units)
9. Industrial Hydraulics -Basic Level (In the English Units)
10. Hydraulic Fluids
11. Hydraulic Filters: Construction, Installation Locations, and Specifications
12. Hydraulic Power Packs (In the SI Units)
13. Power Packs in Hydraulic Systems (In the English Units)
14. Hydraulic Cylinders (In the SI Units)
15. Hydraulic Linear Actuators (In the English Units)
16. Hydraulic Motors (In the SI Units)
17. Hydraulic Rotary Actuators (In the English Units)
18. Hydraulic Accumulators and Circuits (In the SI Units)
19. Accumulators in Hydraulic Systems (In the English Units)
20. Hydraulic Pipes, Tubes, and Hoses (In the SI Units)
21. Pipes, Tubes, and Hoses in Hydraulic Systems (In the English Units)
22. Design of Industrial Hydraulic Systems (In the SI Units)
23. Design Concepts in Industrial Hydraulic Systems (In the English Units)
24. Maintenance, Troubleshooting, and Safety in Hydraulic Systems
25. Hydrostatic Transmissions (HSTs) (In the SI Units)
26. Concepts of Hydrostatic Transmissions (In the English Units)
27. Load Sensing Hydraulic Systems (In the SI Units)

28. Concepts of Load Sensing Hydraulic Systems (In the English Units)

29. Electro-hydraulic Proportional Valves

30. Electro-hydraulic Servo Valves

31. Cartridge Valves

32. Electro-hydraulic Systems and Relay Circuits

33. Practical Book: Pneumatics - Basic Level

34. Practical Book: Electro-pneumatics - Basic Level

35. Practical Book: Industrial Hydraulics – Basic Level

36. Programmable Logic Controllers and Programming Concepts

37. Compressed Air Dryers

38. Hydraulic Circuits – Identification of Components and Analysis

For more details, please visit: **https://jojibooks.com**

About the Author

Joji Parambath is a trainer in the field of Pneumatics, Hydraulics, and PLC for over 25 years. During his career, he has trained numerous professionals from the industries as well as faculty members and students of engineering institutions.

Currently, he is the lead trainer at Fluidsys Training Centre in Bangalore, India (https://fluidsys.org), which provides training in Pneumatics and Hydraulics. He has already written two books on Pneumatics and Hydraulics. The publication of this 32-book series is intended to restructure and update the existing books.

The author thanks all trainees for their lively interaction and many useful suggestions during the training programs, which prompted the author to write this series of books. You may send your feedback to joji.p@hotmail.com.

10th June 2020